THE DRY EXOTIC GARDEN

A GUIDE TO XERISCAPING WITH SUCCULENTS

Paul Spracklin

THE DRY EXOTIC GARDEN

A GUIDE TO XERISCAPING WITH SUCCULENTS

THE CROWOOD PRESS

CONTENTS

FOREWORD

By Robert Brett, Curator of RHS Garden Hyde Hall

I have always been a lover of exotic houseplants, particularly cacti and succulents. Perhaps it is their exotic nature or unusual look, which is so different from the usual array of plants found in our temperate climate. However, over the years, I have been fortunate to travel and visit many gardens, where cacti and succulents were an instrumental part of an outdoor garden design. Of course, the climate and conditions were favourable for xeriscaping, but dry gardens in Arizona and Morocco, as well as the glasshouses in the UK, have all inspired me to 'push the boundaries of horticulture' when I moved to my present role at RHS Garden Hyde Hall in 2015.

Climatically, RHS Garden Hyde Hall sits in one of the driest regions in the UK with an average rainfall of less than Beirut. Matthew Wilson, one of my predecessors, started the Dry Garden at Hyde Hall, with its Mediterranean style and arid planting, in 2001. And over a decade later, when I arrived, I found plants growing outside that I was only accustomed to seeing in warmer climes or inside glasshouses. Whether it is the American agave, the African aloe, or the *Echium pininana* from the Canary Islands with its upright columnar drifts flowering in spring, this garden never fails to impress. I know I had never seen the likes of it before, except in the Canary Islands or at Chelsea Physic Garden, with its protected walled garden.

Soon after arriving at Hyde Hall, I met Paul Spracklin, one of the volunteers, and marvelled at his considerable knowledge of xerophytic plants. Paul's passion, drive, commitment and enthusiasm can be seen in his garden. When I was invited to visit, I was taken back to my memories of Le Jardin Majorelle, Sonora Desert Museum, Huntingdon Botanic Gardens and numerous other places and the words, 'What? You grow that here!' parted from my lips on many occasions. Not only that, but Paul had been doing it for some 26 years, showing how, given the right conditions, cacti and succulents can thrive in an outdoor setting.

I am in awe of Paul's achievement in his own garden, but I am also a great believer in challenging the norms. Climate change is with us, we are predicting a 3–4°C rise in temperature globally and it means that we are going to have to challenge our views and stereotypes of what plants we should be growing for the future. The South-East and Essex could see similar climatic conditions to Barcelona. It is with such perspective and intent that we can utilise Paul Spracklin's wealth of experience and expertise to help create our own xerophytic gardens and areas, outside and all year round, as we have done at Hyde Hall. We pushed the boundaries with the Dry Garden, let's see what else we can do. And who knows, maybe some of the houseplants of the past will become our garden plants for the future.

LEFT: RHS Hyde Hall.

OPPOSITE PAGE: Cacti and succulents in the late afternoon sun.

PREFACE

Exotic or subtropical gardening is currently enjoying a surge in popularity. Certainly, with climate-change models predicting milder winters and hotter summers, growing plants from warmer climates has never been more appropriate. Scientists also warn us that rainfall patterns are set to change with very little rain falling when our gardens need it the most: there is the very real possibility that hosepipe restrictions may end up being the rule, not the exception. So, what then of our tree ferns and bananas?

Necessity forced me to confront this problem. As a devotee of exotic gardening for nearly 40 years, I was first seduced by the prospect of turning my back garden into a jungle. But I live in coastal Essex where, on average, we can expect just 50cm of rainfall annually. Years of struggling to keep my luscious big leaves supplied with the copious amounts of water they demand – not to mention high water bills – made me seek out plants more suited to my arid summer-growing conditions. And I found succulents. Imposing succulent plants with strong architectural form that, grown together, still paint the exotic picture I craved, yet could laugh at the hosepipe ban. A new word entered my everyday vocabulary – xeriscaping.

The opportunity to expand and develop a larger garden from scratch, with xeriscaping foremost in mind, came along in 1994 when I moved to my current house with its, then, one-third-acre (1.350sq m) garden. For the exotic twist it was necessary to take the concept to extremes: cacti and other succulent plants are well known for tolerating excessive heat and drought, but it is only by modifying the drainage of the growing medium to the extreme that we can optimise their chance of surviving our cool, wet winters.

It has become an all-consuming quest: 30 years of painstaking research and old-fashioned trial and error to find a palette of plants that revel in hot, arid summers but, crucially, withstand cold and wet winters. This journey has taken me from the suburbs of Essex to the mountains of north-east and central Mexico and I have drawn upon these experiences to form the basis of this book. Whether contemplating a new design for a complete garden, a makeover for a sunny corner or simply a few containers for the patio, xeriscaping represents a viable option for future-proofing our gardens.

LEFT: The main succulent bank in my own garden enjoying the drainage provided by a steep slope.

OPPOSITE PAGE: The iconic Dry Garden at RHS Hyde Hall is perfectly in tune with the low rainfall in this part of Essex.

PART I

XERISCAPING AND YOUR GARDEN

Xeriscaping is essentially landscaping to minimise the need for irrigation. The word itself is largely self-explanatory: a catchy portmanteau that combines the word 'landscape' with the Greek prefix 'xero-', meaning dry. The term was coined originally in the USA in the early 1980s and, although the concept is nothing new, the actual term 'xeriscaping' somehow seems to encapsulate how to manage your garden landscape sustainably during extended dry periods.

The most effective way to achieve this is to choose plants that are adapted to coping with seasonal drought and it is gardeners in the county of Essex – the driest in the UK – that have been leading the way. Beth Chatto, in her groundbreaking 1978 book *The Dry Garden*, caused a paradigm shift in how people garden in this country with her maxim of 'right plant, right place'. This laid the foundation for the iconic Dry Garden at RHS Hyde Hall, which was completed in 2001. Matthew Wilson, curator at the time, and the team expanded on the idea with an initial layer of rubble underneath the soil/grit growing medium to ensure keen drainage and to add topographical interest to an otherwise even slope. Careful species selection then gave rise to another maxim, 'plant wisely, water once', as only when initially set into place are plants ever given water.

CHAPTER 1

SUCCULENTS AND CLIMATE-ADAPTIVE PLANTING

Cacti and succulents do not all live in deserts; in fact, not many do. Some of the succulent plants that I grow would see more rain in their original habitat than they do here in my Essex garden. The rainfall pattern is different, however: they might enjoy a wet spring and early summer but then have to endure months of dry weather during the rest of summer and into winter. To tide them over these dry periods, succulents have evolved a variety of strategies, principal among these being the development of tissues for storing water, in their leaves, stems or roots, which is taken up in rainy spells.

Fleshy leaves on this *Aloe brevifolia* are adapted for water storage.

Having accumulated this supply there are then a number of methods employed to mitigate water loss in their day-to-day affairs. Generally, plants lose most of their water through their leaves, so modifying these is a first important step. The fleshy, water-storing leaves of many succulents have developed a thick skin, often waxy, that slows down evaporation. Cacti address this by ditching leaves almost entirely – reducing them to spines – while the fat, green, water-filled stems assume the role of photosynthesis and other leaf-related activities.

Cactus spines are, in fact, modified and reduced leaves.

Crassulacean Acid Metabolism (CAM)

An important part of plant biology is how they exchange gases with the air via microscopic pores on the leaf surface called stomata. It is necessary for these pores to open periodically to facilitate the movement of oxygen and carbon dioxide in and out of the plant but, whilst open, there is an accompanying loss of water vapour from inside the plant. In most plants this process takes place during the day when water vapour loss is at its highest. A few groups of succulent plants, however, have evolved biochemical processes that allow them to only open their stomata at night to carry out gas exchanging, effectively holding their breath all day. This is known as Crassulacean Acid Metabolism, or CAM, and is believed to be a major contributory factor to the success of these plants.

Representation of stomata set within cells of a leaf surface.

Succulents and Cold

Frost and snow – these words can strike terror into the heart of an exotic gardener and often induce a mild sense of paranoia. Will this year be 'The Year' when a polar continental air mass sweeps across from Siberia? Fortunately such events are rare with climate-change models indicating a shift towards milder, wetter winters. Nevertheless, to a greater or lesser degree, frost and snow are regular occurrences across the UK and need to be considered.

Frost

Frost is a broad term for a number of slightly different weather phenomena. In winter we generally get air frosts, when the temperature of the air falls below freezing, causing ice to form on surfaces. A ground frost happens most often in spring and autumn when clear skies allow heat from the ground to radiate away and the temperature of exposed surfaces to dip below freezing, even if the air temperature remains above freezing. A hoar frost is when dew forms on surfaces that are already below freezing, causing beautiful feathery ice crystals to form. A white frost is when dew forms and the temperature subsequently falls below freezing, leaving a uniform coating of white ice. A lot of succulent plants see frosts of all types in their habitat and so, largely, it should not be a primary concern, especially if species have been chosen carefully.

The beautiful feathery ice crystals of hoar frost coating the leaves of this dasylirion.

A white frost beautifying these cactus spines with ice.

Snow

For the succulent garden, snow is by far the most damaging winter event. It is often stated that the Inuit have 50 words for snow but, here in the UK, it is really only necessary to make a distinction between two types – wet and dry. Because of our temperate maritime climate, we generally get wet snow. Wet snow can fall even when temperatures are above freezing and it contains two to three times the amount of water than dry snow. This makes it heavy, causing

Melting snow on this euphorbia dripping into the nooks and crannies of the plant.

Wet snow on this agave has partially thawed then frozen again.

Wrong Type of Snow

The difference between wet and dry snow in the UK was brought into focus in February 1991 when rail services ground to a halt due to a rare fall of deep, dry snow – the track-clearing machinery, being designed to deal with wet snow, was unable to cope. The expression 'wrong type of snow' was splashed all over the media and 'snowclones' have now entered the language – 'wrong type of leaves', 'wrong type of seaweed' and so on. For growers of succulent plants, it is very much the right type of snow!

mechanical damage, and leads to cycles of freeze and thaw that can penetrate right into the heart of a plant.

Ironically, perhaps, in very cold climates the snow is generally dry snow – formed when precipitation occurs during periods that are significantly below freezing – and this dry, light, powdery snow actually insulates plants against the effects of severe freezing. Thus even if, in their habitat, the succulents experience deep freezes, it is usually accompanied by dry weather or a layer of insulating powdery snow. An example of this would include many forms of *Agave utahensis* and this is part of the reason why they fail here despite being exceptionally cold-tolerant.

Timing

The length, depth and timing of cold can make a massive difference. Some plants can withstand a brief dip down to, say, –10°C, if it is only for an hour or two on one night and the daytime temperatures the following day rise above freezing. The same plant might die if exposed to prolonged freezes of –6°C over several nights with daytime temperatures staying below freezing for several days.

One of the most critical things, however, is the timing of when the cold arrives. Throughout this book, to save repeating facts and figures every time, I will make references to my 'worst winter', 'late cold' winter and 'early cold' winter. These have been my benchmark winters, each of which has taught me valuable lessons in how plants behave.

My 'worst winter' was the winter of 2009–10, during which time wet snow arrived on 17 December, accumulated through the winter to end up 60cm deep, and stayed on the ground until April, constantly cycling between freeze and thaw. The absolute low temperature was one night in January when it reached –8°C, and there was a fortnight when the temperature did not rise above freezing day or night. I had started to plant my new garden in earnest during 1996 and each of the subsequent thirteen winters had been milder than average. This led me to become complacent and increasingly reckless in my plantings, reasoning that plant X had survived,

A heavy snowfall during my worst winter.

so therefore I should push things a little further and plant Y and Z, too. In that regard, winter 2009–10 was a much needed wake-up call. A lot of plants survived, however, as they had already gone into dormancy during a chilly autumn.

The 'late cold' winter was the weather event in 2018 dubbed by the press 'The Beast from the East' but more formally 'Anti-Cyclone Hartmut'. After one or two frosts in early November, the following December, January and most of February were particularly mild. By mid-February light levels had improved and the mild temperatures had encouraged a lot of plants to break dormancy and start into growth. Then, suddenly, on 22 February a dreaded polar continental air mass moved in from Siberia bringing wet snow and falling temperatures down to –6°C. Tender new growth was suddenly exposed to cold and the results were catastrophic: plants that would have easily taken those conditions when dormant had been damaged or killed.

The 'early cold' winter was December 2022 and, in some ways, the most damaging. Snow arrived on 10 December accompanied by a freeze down to –6°C and a period of eight days when temperatures struggled to lift above freezing. In themselves these figures should not have been too much of a problem but, again, it was a question of really bad timing: a confluence of extreme events that, hopefully, will not be repeated frequently. We had just experienced an unprecedentedly long, hot and dry summer, and many plants, even some succulents, had been stressed by this and gone into dormancy. That was followed by an astonishingly wet and mild autumn, which broke the dormancy of these plants: an awful lot of them were still actively growing coming into December as there was almost a second spring following the drought. Then, suddenly, they were subjected to this rapid downturn in temperature with a wet snowfall that went on to freeze/thaw for a whole week. Some of the worst hits were actually plants that I had eagerly sought out – those from more temperate regions or high altitudes. Not requiring as much heat, they were still happily growing when the cold arrived, whereas some of the more xeric and heat-demanding plants had already started to go dormant and, therefore, got off more lightly.

Cold Damage

Damage caused by cold can manifest itself in many ways but the reason why it happens is essentially the same. The contents of the plants' cells freeze and expand, bursting the cell walls; there is no way back from this and the cells die. Depending on the severity of the cold/tenderness of the plant, this can be just localised or it can affect the whole plant. Either way it only becomes apparent once the freeze has passed and the inside of the plant has thawed out. If the damage is confined to small areas this shows up as spotting, crazing or splitting of the leaf. If the whole plant has been affected it will lose its structural integrity and just collapse. Something that has happened occasionally to agaves in my garden, when they have been exposed to a long freeze/thaw cycle, is that they seem to survive just fine but then, once temperatures recover in spring, the middle suddenly falls away from the base of the plant, taking the main growth point along with it.

Offsets may or may not be affected, so it is usually wise to keep damaged plants *in situ* and see what happens. Some plants, such as beschornerias and trunkless nolinas, have a large underground caudex from which they will almost certainly put up new growth. Patience is particularly recommended for some

Cold damage on this aloe has caused it to collapse, now resembling a heap of boiled spinach.

Damage to this agave became evident in late spring.

Cold damage to the stems of *Aloiampelos* (*Aloe*) *striatula*.

Signs of recovery in early May.

stemmed succulent plants. In the case of *Aloiampelos* (*Aloe*) *striatula* the soft top-growth starts to show damage below around –6°C, but if the woodier stems underneath are still firm, new growth will sprout from these in spring. Flowers are borne on the previous season's growth so are lost for that year but the plant recovers. Even if the woody stems are lost, which starts to happen at around –10°C, it is possible that new shoots will emerge from underground in spring. A friend has documented that his established specimen came back from underground after a staggeringly low –18°C in the mid-1980s, a winter that saw off many of his 'hardy' garden plants.

The effects of frost and cold can be mitigated in different ways – by some form of overhead protection, insulation, careful use of microclimates but, largely, by judicious species choice. Much of this is discussed in the coming chapters.

Hardiness

'Is this plant hardy?' A relatively straightforward and reasonable question, you would think. If only there was a straightforward and reasonable answer! Whether a plant is going to be hardy or not for any given location depends upon an extremely complicated cocktail of interdependent factors.

Meaning

What does 'hardy' actually mean? It certainly means different things to different people, although a statement along the lines of 'able to withstand year-round climatic conditions, including frost, without protection' seems a reasonable consensus. But evaluated over what period – and do we include the record-breaking winters or just average years? For some people a plant that scrapes by in a mild winter, with some protection, is deemed to be 'hardy'. Conversely, some may only consider a plant to be hardy if it can survive the lowest historically recorded temperature in its location. Most of us, I think, have such things loosely compartmentalised in our mind. The tricky bit is to understand what someone else means when they say, 'Oh, yes, this plant is perfectly hardy' because, as with any review, knowledge of the reviewer is the only way to make sense of it.

Hardiness Ratings

Various attempts have been made to categorise either how cold-hardy plants are, or how cold geographical locations get, by 'ratings' or 'zones'. None give a complete picture; at best they can be used as a very rough guide, at worst they are downright misleading.

The USDA Zone system seems to have been almost universally adopted, yet, to my mind, it is the least appropriate for the UK. It was devised by the US Department of Agriculture to describe all parts of America by taking the average annual minimum temperature reached there. These temperatures are divided up into ranges or bands, which are then transcribed on to a map, rather like contours, linking areas that experience similar figures for the mean low temperatures each winter. So, from this table, it can be seen that for a plant to be considered hardy in USDA Zone 7, it must be able to withstand temperatures between 0°F and 10°F, which translates clumsily to –17.8°C and –12.3°C.

The stumbling block is that it is based on a single parameter – how cold it gets in winter – and nothing else. It is purely a number-crunching statistic. It is simply not possible to relate climate in the UK to climate in the USA. To use an extreme example, Gainesville in Florida is rated to be in USDA Zone 9a as are the Shetland Isles off the coast of Scotland because they share historical winter low temperatures. Clearly the same plants will not thrive in both, as Gainesville is about 20°C hotter in summer!

The RHS introduced a table of hardiness ratings after lengthy collaboration between horticultural professionals, which, I believe, is more relevant when discussing plants grown in the UK. This system pertains to plants – not places – and their winter performance when grown in the UK. So a plant described as H5 would be hardy to between –10°C and –15°C. It is still not a perfect system, still takes no account of conditions during the growing season, but is, I feel, much more appropriate and is the one I use here.

USDA	Temperature °F
5b	–10 to –15
6a	–10 to –5
6b	–5 to 0
7a	0 to 5
7b	5 to 10
8a	10 to 15
8b	15 to 20
9a	20 to 25
9b	25 to 30
10a	30 to 35
10b	35 to 40
11a	40 to 45

RHS	Temperature °C
H7	–20 and below
H6	–15 to –20
H5	–10 to –15
H4	–5 to –10
H3	0 to –5
H2	5 to 0
H1c	10 to 5
H1b	10 to 15
H1a	Over 15

Diagram comparing USDA and RHS hardiness zones.

Location, Location, Location

Considering that the UK is such a small archipelago, the range of climates is quite staggering. Broadly speaking we enjoy a cool-temperate, oceanic climate, benefitting from being bathed by the warm currents of the North Atlantic Drift and other, possibly even more significant, influences. This is one reason why London, despite being at the same latitude as parts of Alaska, is so much milder. Within these isles, however, there is a marked gradation in rainfall, summer warmth and winter cold. It is not a simple north/south divide but a composite puzzle.

On any group of islands it is important to note that there is a temperature gradient moving from coast to inland. Bodies of water lose heat accumulated over the summer much more

Agave americana 'Mediopicta Alba' in a seafront garden. Even a couple of kilometres can make a huge difference.

A bed at the Minack Theatre in south-west Cornwall, where aeoniums and more tender winter-growing plants thrive.

slowly than landmasses, and anywhere near a coastline will, therefore, have cold mitigated to some extent by proximity to the warmer water. For example, about 7km east from my home, and right on the south Essex seafront at Westcliff-on-Sea, there is an *Agave americana* 'Mediopicta Alba' growing quite happily in the garden of some apartments, a plant I have struggled to keep alive just a little further inland.

Generally speaking, the west side of the UK experiences wetter, cooler summers and wetter, milder winters due to the influence of the prevalent south-westerly weather systems. Conversely, the east side is drier and sunnier in summer, with more influence from nearby continental Europe, but potentially colder in winter. Simple physics would imply that winters should be colder as you move further north but in practice the situation is far more complex: the west side of Scotland, for example, sees a strong effect from warm water currents and is actually extremely mild in winter. Dotted about are the large urban conurbations giving rise to little heat islands of buildings and tarmac: London springs to mind, of course, but also Manchester and Birmingham. It is so confusing.

All of this means that there is no 'one size fits all' or 'buy it off the peg' answer: a different portfolio of plants needs to be considered according to where you live. But, with absolute certainty, one thing it would be fair to say is that wherever you live in the UK, there will be cacti and succulents available for you to grow outside.

In the extremely dry south-east corner of Essex, my garden is suited to growing cacti.

Stuff Happens

You will lose plants. Every winter is different, every location is different, plants will inevitably behave differently and you cannot conceivably think of everything every time. The overriding principle here is to remain philosophical whatever the winters bring. Mourn the losses, but do not become disheartened. Think about what you might realistically have been able to do to change the outcome. Would the plant have survived if kept dry? Consider trying again with a cover. If you do not want to cover, but the winter was exceptional, try it once more but this time in a slightly different spot. Perhaps at the foot of a bigger plant that offers some degree of shelter. If not, draw a line underneath and move on to something else. That is the nature of experiments: you always get results, even if they are not necessarily the ones for which you hoped. Do not dwell on the plant you have lost but think of the opportunity that has now arisen to try something new that is just as exciting. You can rejoice in the successes, especially the unexpected ones. In that respect it is the punishing winters that teach us more than the mild ones, and that hard-earned knowledge is precious. Xeriscaping in the UK is still very much uncharted territory. Our questions may not have answers; indeed, we may not even know the questions yet. It is up to us to find out.

The heart-lifting sight of strong regrowth on *Schinus molle*.

Aloiampelos striatula at RHS Hyde Hall making a strong recovery by October following defoliation.

CHAPTER 2

PRACTICAL CONSIDERATIONS

Provenance

When talking about plants, their provenance means the 'place of origin' or habitat which, in many cases, can be a clue as to whether a plant will be hardy or not. As a general rule, plants that come from a similar climate to that in which you are trying to grow them will be hardy. Conversely, those from a much warmer climate, you would not expect to be hardy. Of course there are always exceptions. *Agave stricta*, for example, would never, ever see a frost in its southern Mexico habitat but has survived –8°C in my garden unblemished.

Things to consider when deciding which species to try would include the latitude at which they grow, the altitude and also whether they grow in a rain shadow or not. Using *Agave montana* as an example, this beautiful plant comes from around 3,100m altitude in the north-east of Mexico just on the very edge of the Tropic of Cancer and on the damp side of a mountain system. Nights, even in summer, can be misty and chilly. In winter there is often snow lying on the ground. As a consequence it is beginning to establish itself as a successful succulent throughout much of the UK, although it does seem to resent the type of wet snow we get here and is best with a cover in winter.

Provenance can be especially helpful when considering a plant that has an extremely wide distribution in the wild. Trials of eucalyptus and acacia trees have revealed that plants raised from seed collected from populations at the coldest edges of the species' natural range exhibited more cold tolerance than plants raised from subtropical provenances. They might be the same species but not all can be grown successfully in the UK. This could, theoretically, apply to some succulent plants, e.g. *Aristaloe aristata*, the natural range of which extends across some very high peaks in the Drakensberg mountains in southern Africa. Plants grown from seed collected from a known, high-altitude provenance would be very desirable.

OPPOSITE PAGE: *Agave montana* in its mountain habitat.

Sunlight

When grown under glass, succulents will sometimes scorch in full sun but this does not happen when they are planted outside – fresh, moving air prevents this occurring. After all, succulent plants mainly come from hot, sunny countries, so a bit of British sunshine should not be a problem for them. Cacti and succulents really do grow best when given as much sun as possible: they become more compact, colours are more pronounced, spines are stronger; growth is healthier, hardier, faster. Some years ago I gave a friend of mine a spare yucca seedling, retaining one myself. We planted them both out the same year, treating our plants with similar benign neglect. Mine

Grown in full sun the spines of this cactus are stronger, the colour more intense.

was in a position where it received maybe five hours of sunshine each day in summer, whereas his was in full sun from dawn until dusk. He lived nearby so the comparison was interesting, if unintended. Five years on, my plant had grown into quite a nice little thing, maybe 40cm high, but with no sign of a trunk. By contrast my friend's plant was over 1m tall with a clear trunk and far more robust and impressive in every way.

There are also unseen benefits from growing plants in full sun. We often hear of hot summers 'ripening growth' but what does that mean and how can it affect a plant's hardiness? Many of our ornamental woody trees and shrubs come from regions of the world that have long, hot summers. Even if the winters are punishingly cold, a long, hot growing season means that the tender new shoots made in spring have a greater opportunity to mature, largely through changes in carbohydrate levels, into tougher woody stems that better prepare them to cope with winter. This is apparent in this country as there is, in fact, a marked difference as you go from the south-eastern tip of England and move north and west in the types of plants that thrive – many only perform best in the hotter south-east, some prefer the cooler north-west.

A similar principle can also be applied to succulent plants. Full sun and higher temperatures will often induce changes to the internal chemical composition of the sap within plant cells that effectively turn their contents into 'antifreeze'. This enables them to withstand far lower winter temperatures than they can in a more temperate climate: winters may be milder but they just do not get enough warmth in summer to change internally. This is particularly relevant when looking at cold-hardy cacti and succulent references from the USA. It is so tempting to splash out on a purported cold-hardy plant rated suitable for USDA Z6 or below only to see it fail miserably in the UK's milder winter climate because it lacked the long summer baking. Most forms of *Agave utahensis* would fall into this category. So, again, it follows that best practice is to grow our succulent plants in the sunniest, warmest spot available to give them their best chance.

Agave utahensis enjoying a brief life in the sun.

Drainage

Many of our heat-loving desert plants are extremely hardy to cold but if they sit with their 'feet in water' they will rot and die. So it really is particularly important to pay attention to drainage. In fact I would say this is the single most important factor for success. The underlying principle is that it is easier to give a plant water if it needs it than remove it when it does not.

A rather confused frog, wondering where the water has gone.

All my succulent plants are growing in what are effectively raised beds, filled with material to ensure free drainage.

Dotted around various parts of the country are little pockets of gravelly, flinty soils – commonly encountered in the north of Essex and parts of Norfolk, for example. If you are lucky enough to be so blessed you can turn what is perceived as a real problem in 'standard' gardening circles into a massive advantage when it comes to growing succulents, with very little soil amendment needed. If not, you have to look at other ways forward and, for most intents and purposes, this means using a raised bed of some kind.

Raised Beds

There are many ways to make a raised bed: choices might depend upon what materials are available locally or how the appearance of the planned bed sits within the design of the wider garden. Natural stone is usually an excellent, if expensive, material to use, being heavy and stable, managing to look just right in so many situations, whether in a naturalistic rockery setting or a more modern design. If it is quarried locally, so much the better, as it will sit more happily in the landscape and reduce costs. Being devoid of natural rock deposits here in Essex, I decided to look further afield when sourcing stone for my large rock garden, ultimately settling on lava rock imported

A 'small cliff' of lava rock in my garden just after construction and planting.

Gabions used to great effect in Chris Grundy's London garden, the border appearing to vanish under a path only to re-emerge on the other side.

Hollow concrete blocks, broken and pieced back together, make an interesting fill for gabions in James Knock's garden near Brentwood.

Timber sleepers utilised for a new project in James Tucker's Cambridgeshire garden.

Plastic drainage pipes repurposed as a low retaining wall.

Screened furnace bottom ash makes the perfect backfill material.

from Turkey. This fitted so well with the exotic concept, adding to the deception that the garden was somewhere hot and sunny. The initial cost was high but the value added to the garden has been immeasurable.

Originally the preserve of road cuts and railway embankments, gabions are being utilised increasingly in domestic situations and offer amazing flexibility in terms of how they can be shaped and, especially, filled. Gabion grade rock is available but also recycled materials, such as crushed concrete, crushed brick and general rubble, all work just as well. Timber sleepers are a quick and easy way to raise a border with modern, clean, treated hardwood versions replacing the tar-soaked baulks recycled from the rail tracks of Europe.

Often, however, it is a good opportunity to think outside the box and use something a bit unusual. In one section of my garden I have cut lengths of large diameter ribbed plastic drainage pipes, usually seen alongside new road construction sites, bolted together with stainless steel fixings and placed upright as a retaining wall – the inside of the pipes filled with rubble and aggregate and then planted.

Drainage Material

The choice of drainage material is as divisive as it is diverse with many gardeners and growers dogmatically entrenched in their beliefs about what are, or are not, suitable media in which to grow plants. I can only speak from my own hands-on experience. Over the years I have moved away from mixing my native clay soil with freely draining material towards a completely inorganic substrate. No soil or organic matter at all, as I believe it holds on to moisture far too much. When filling a raised bed I start by using chunky rubble of broken bricks and concrete to fill the larger spaces, blinding these with some form of aggregate, planting as I go. I find that the plants enjoy wrapping their roots around these big chunks and it can only aid their stability. For the aggregate, at one point I was able to get hold of screened 'furnace bottom ash' – which goes by various common terms such as slag or clinker – a waste product from coal-fired power stations. It is increasingly unavailable due to decommissioning of these power stations, any stockpiles of material being used to make lightweight concrete blocks. In some former coal-mining areas of the UK, however, it might still be possible to obtain, so it is worth checking. I found it to be perfect. Other fancier materials I have used include LECA (lightweight expanded clay aggregate) and moler clay, available as granules for mopping up petrol spills and even as a low-dust, non-clumping cat litter! These days, however, I mainly just use all-in ballast, which is a mixture of 20mm gravel and sharp sand. It is easily and cheaply available from most builders' merchants or DIY sheds and I have had no detrimental issues when using it. This all seems very cruel but I have found that plants grown really hard like this are tough, compact, have nice form and are better placed to deal with winter weather.

Winter Protection

Having settled that winter rain and snow are the nemeses of cacti and succulents, one of the most effective ways of dealing with this is to cover them somehow to keep the weather off. There are so many ways of going about this, from the elaborate and extravagant to just the simple expediency of placing a layer of fleece over a plant when it snows, then removing it once it stops.

Various acquaintances of mine adopt different methods. On the west side of Norwich, Norfolk, Melissa Scott and her partner Keith construct a complex timber and polycarbonate township over their entire succulent borders, tailor-made to the contours, erected and dismantled over the course of a fortnight or so at each end of the winter. Their glorious garden, Crete Lodge, is a testament to the effectiveness of their painstaking efforts.

Darren Hobbs in Newport, South Wales, has a number of choice and unusual plants that he goes to some lengths to protect. Many cycads, including a magnificent specimen of

Macrozamia communis, enjoy a sleeping bag slipped over the leaves if the weather threatens to get much below –5°C. I would say his crowning achievement, however, has been to keep a specimen of *Aloidendron* 'Hercules' alive since 2011 by building an increasingly larger temporary greenhouse around it. This magnificent giant now towers over 3m and there is not another plant like it anywhere in the UK.

Tom Hart Dyke, of World Garden fame at Lullingstone Castle, Kent, erects a large polytunnel over 'Mexico'. The effectiveness of this we will see shortly. There is a similar story for Ali and Si, whose former garden near Manningtree in North Essex contained plants I would struggle to grow in the open, such as the stunning *Agave weberi*, yet saw winter low temperatures way below mine – a polytunnel over the entire arid border allowed this to happen. Away from the south, in fact way up in the West Yorkshire Pennines, Joel Robinson manages to grow a vast array of cacti and succulent plants, covering his plants during the wet winters there.

My own choice is to end up playing chicken with the weather. I have bits and pieces on hand to erect some adapted

A winter scene at Crete Lodge, Norfolk.

Aloidendron 'Hercules' snug and dry in its Welsh winter quarters.

Agave weberi happy in north Essex with a lean-to polytunnel for protection.

Framework in place ready for the cover for my cactus bed.

The same border kept dry during a snowfall.

polytunnels quickly if things start to look a bit dicey. Also a selection of plastic umbrella cloches, clear umbrellas from 'Pound Shop' type outlets, rigid polycarbonate domes, a big roll of horticultural fleece, fleece blankets, bobble hats and sleeping bags from charity shops – all of which come in extremely handy if needed. For the most part, however, I leave well alone, preferring to keep my plants open to the air – although I do spend most of the winter anxiously fretting over forecasts.

A clear umbrella cloche keeps snow from penetrating the heart of succulents.

A small polycarbonate cloche placed on top of a columnar cactus keeps the growing tip protected.

Charity shop bobble hats can be life savers and give neighbours something to discuss.

Simply laying a fleece blanket over a vulnerable plant can make a difference.

Why Cover?

The massive advantage of having a winter cover is that you can almost exponentially increase the range and diversity of plants you grow by keeping the area dry, especially if you get the covers in place before the weather turns wet. This was brought into sharp relief for me over the winter of 2009/10, which was the coldest winter I have had since I started gardening here in Essex. A policy decision on my part at that time was that I was not going to cover any plants at all. As detailed earlier, in the open garden I recorded a low temperature of –8°C, a period of two weeks where the temperature remained below freezing and 60cm of wet slushy snow that fell in December and lay on the ground, going through freeze/thaw cycles, until finally disappearing in April. I lost many, many plants that winter, amongst which was a group of three *Agave mitis*. The following year, about 48km (30 miles) south-west as the crow flies, Tom Hart Dyke at Lullingstone endured a low of a staggering –16.9°C yet, under his dry Mexican polytunnel, the *Agave mitis* were just fine. Suffice to say, my policy decision has now been amended somewhat.

Agave mitis doing an impersonation of boiled spinach after my worst winter.

A large polytunnel erected over 'Mexico' at The World Garden, Lullingstone.

Air Circulation

This is something that largely gets overlooked but I believe it is actually quite important with many benefits. As mentioned above, fresh, moving air prevents plants in full sun from scorching in summer. During cooler, wetter periods, stagnant humid air can allow the build-up of bacteria, fungal spores and general pathogens, which, in turn, can lead to all sorts of rotting, so a healthy air movement helps to prevent this. To that end, I have found it is better not to plant too densely near the more susceptible plants, something that can effectively trap moist air around them. There is also a beneficial 'wicking' effect whereby humid air is swept away from the base of the plant, drying the soil surface. Moisture is then wicked up the soil to the surface to replace it – eventually drying out the soil. A good analogy would be washing pegged on a clothesline drying much more quickly on a windy day. This all becomes particularly relevant in winter if you choose to cover your plants – if possible, keep any plastic covers ventilated.

Propagation

Succulent plants conveniently offer the grower a variety of propagation methods.

Seed

Growing from seed is an excellent and inexpensive way to obtain more plants. For some of the rarer plants you may choose to grow, it could well be the only way if they are not available commercially other than seed. Even if it is a commonly grown plant, however, there is always the possibility that, with a shake of the genetic dice, seedling variation will throw up a chance plant that is a bit different from usual – stronger teeth, more compact. You never quite know.

Seed sowing is another area of horticulture where methods and opinions are varied and divided. Some growers state categorically that compost needs to be sterilised before sowing seed, leading to robust discussion over the best method to do so: microwave or pouring boiling water over the pot seem popular. Others, me included, do not bother. Seeds of the woody lilies, such as agaves, are generally quite large and best covered by 2 or 3mm of medium. For this I tend to use some form of heavier grit that keeps the little seedlings weighed down, as once they start to germinate they will often push themselves out of the medium rather than root down into it. Many succulents, such as echeverias, have seeds that are dust fine and best sown on to the surface and left uncovered.

Either way, enclosing the pot somehow, once the seed is sown, keeps the moisture and humidity levels high, which aids germination. Placing the pot in a Ziploc bag is often recommended: if I have any to hand I sometimes use yoghurt

Agave seed sown on top of moistened moler clay.

Three months later, little agaves crowd the pot where they will remain until the following spring.

The seed is then covered with a 3mm layer of LECA.

pots with clear, tight-fitting lids. Once the seed has germinated, this closed atmosphere is best removed by opening the plastic bag or taking the lid off, whilst at the same time keeping humidity levels quite high until the little plants are a few months old. Some growers bypass a lot of these early pitfalls and start seeds off by scattering them on to damp tissue paper. I recently saw some finessing of this technique whereby the tissue was kept damp using hydrogen peroxide solution, reducing the incidence of fungal attack to a minimum. Then, once germinated, the delicate little things are carefully transplanted directly into individual pots. I have not tried this myself but have read good reports. Whichever method you employ, make sure you label the pot!

Stem Cuttings

Many succulent plants that have stems are most easily propagated from stem cuttings. Usually the tip but sometimes stem sections seem to work just fine. Using a sharp, sterile blade, cut the stem just below a leaf and leave the cutting for a few days so that the exposed cut area can dry off and callus over before being pushed into the growing medium. I have, on occasion, taken cuttings from my *Aloiampelos striatula* then left them, forgotten, on a bench somewhere only to stumble upon them a few weeks later at which point they have usually already started to push roots out of the stems. It really is that simple, sometimes.

Four months after cutting, these aloe stems are rooting into the air by way of a hint to me.

Offsets

Quite a lot of rosette-forming succulents, e.g. many agave species, will make offsets from runners formed at the base of the plant. Sometimes they appear close to the mother plant, in which case care is needed to detach the baby. Sometimes they can appear some distance away from the mother if planted in the ground, or even pop out of the drainage hole in the bottom of a pot if containerised. These are easily severed and rooted to make new plants, again leaving the cut to dry for a few days before potting up.

Agave offset popping out of a pot drainage hole looking for light.

Leaf Cuttings

A few succulents can be propagated from leaves. A number of *Echeveria* species will make new plants from a basal leaf that is detached, left to dry for a few days, then placed and left on the surface of your potting medium. Little roots will soon sprout from the end that was attached to the stem, followed by the start of a small plant. For some species it is more productive to use the leafy bracts that appear on an inflorescence, which can be detached and rooted in the same way. Many gasterias can also be multiplied from basal leaves although, frustratingly, this method will not work for the closely related aloes.

New plants growing from leaves taken from *Graptopetalum paraguayense* and *Echeveria rosea*.

Bulbils

A number of succulents, in particular many agaves and their kin, produce baby plants on their flowering stalk that are known as bulbils. These can simply be detached, partially buried in the growing medium and rooted. This applies particularly to the genus *Furcraea*, the giant flower spikes of which end up festooned with bulbils, which, in some species, virtually replace seed production.

Bulbils removed from the inflorescence of *Furcraea parmentieri*.

Artificial Means

Some succulent plants, however, remain resolutely barren of offspring and are reluctant to provide convenient means of reproduction until they flower and set seed. In the normal course of events, *Agave montana* grows as a solitary rosette for almost all its life, perhaps making an offset of two as it prepares to flower. Fortunately, seed for this species is generally available and plants can be easily and quickly raised this way. This is not always appropriate or possible, however. We might want to increase the numbers of some hybrids, variegates or distinct cultivars that do not readily offset but the desirable characteristics are not passed down genetically through seed.

Tissue Culture

On a commercial scale, plants are propagated artificially by the process of tissue culture (TC), whereby a few cells, usually from a meristem, are dissected out from the plant and raised in growing media under laboratory conditions. These are then divided into as many plants as required – many thousands can be produced from a single mother plant. For nurseries operating within a limited market this is impractical as there is a large initial cost involved for the laboratory set-up, but it is widespread practice in the USA where there is a huge domestic customer base. In recent years there has been a proliferation of named cultivars and hybrids produced in the USA and spread via TC. Taking *Agave ovatifolia* as an example, grown from seed it is extremely variable in size, degree of spination and leaf texture. Seedlings perceived to have market appeal have been selected, put into TC and released with catchy cultivar names such as 'Vanzie', 'Frosty Blue' plus the variegated 'Orca' and 'Frosted Flecks'.

Agave ovatifolia 'Vanzie' is a distinct form of the species propagated by tissue culture.

Offsets growing from the centre of a cored agave.

In these cases it is sometimes possible to force a plant to multiply artificially. On a small domestic scale a plant can be 'cored': a process that involves carefully destroying the main central growth point in the heart of the plant by cutting or drilling it out. This may then encourage it to make offsets from otherwise dormant side buds when, ordinarily, it would not. Caution and care are bywords here: it is a risky procedure, the outcome of which is far from certain and could well lead to the death of the cored plant, leaving you with neither offspring nor mother. I recall how apprehensive I felt when coring a 'one-off' natural hybrid!

Cacti

Cacti are usually started off from seed or, for clumping or branching plants, offsets or cuttings. Opuntias are easily propagated from individual pads, partially buried or even laid flat on the compost surface. When growing cacti under glass, other methods can be employed. They can often be propagated by grafting – used for difficult-to-grow species that are grafted on to roots of easier host plants. This procedure is not appropriate for growing cacti outside, however, as the area of the graft on the more difficult plants is a point of weakness for the tribulations of outdoor life.

Propagating Media

Whichever of the above methods are employed for propagating your plants, I have always found the same growing medium to be universally effective. Again, like growing media for mature plants, opinions can vary massively, but this is what works for me. I usually start off seeds, cuttings, offsets and bulbils in moler clay. It is sterile, drains perfectly, yet absorbs moisture and gradually releases it – this seems to encourage the production of roots just perfectly. Once roots have been formed, I would add into the water a half-strength liquid balanced feed, working up to full strength once there is a healthy growing plant.

When growing-on seedlings, in particular, but also cuttings and bulbils, I tend to leave them all to grow together in the community pot rather than separate them out early. I do not know if there is any science behind it but they seem to do better crowded together in this way. Certainly, left for an extra year, the roots they have grown are larger and less easily damaged when finally transplanting into individual pots. For dasylirion and nolina seedlings, I actually just pot on the whole community pot, tangled root ball and all, into a larger pot and grow them on for another year or two beyond this. I found that I can keep a pot full of agave seedlings alive in a state of almost suspended animation for many years if lack of space in my greenhouse prevents me from moving them on into larger containers. I am almost ashamed to admit finding a pot of *Agave gentryi* seedlings that, when I checked the label, I discovered had germinated some 18 years earlier, still alive and marking time, needing just the odd splash of water to keep them alive. I am sure seed in a fridge would have lost its viability by then so perhaps it is a valid method after all, even if it appears a little unkind.

RIGHT: Moler clay, my preferred growing medium.

BELOW LEFT: Yucca seedlings, five months after germination, happily crowded together.

BELOW RIGHT: The same group moved into a larger, deeper pot where they will remain for a year or two.

CHAPTER 3

DESIGN CONSIDERATIONS

Basic Design Principles

If I wanted to compose a song I could grab a guitar, bash out a few pleasing chords to sit under a melody and think of some words to put to that melody. And, who knows, it might be quite a nice song. But if I understood music theory I could add harmonies, use augmented chords or possibly introduce different scales. I would know exactly when to introduce the bridge, or when and how to switch from major to minor key to change the mood of the listener. It might become very much more than just a nice song. To create something when understanding how it fits together will usually end in a more satisfying, balanced and harmonious composition, and I would say exactly the same thing applies to designing your garden.

A view from the top of my rock bank soon after completion.

A similar view sixteen years later, showing how plants grow and change the relationship of masses to voids.

OPPOSITE PAGE: A group of silver plants.

Garden design is an unusual mix of three components: art, science and time. Designing a space is art. The planting/horticulture is science. The relationship between the space and the plants alters with the passage of time as they grow. Making the effort to understand these processes can only give you a better garden.

Masses and Voids

In many ways it is best to consider designing a garden to be a two-stage process, starting with how best to use the space – this being the balance between masses and voids, together with movement through it. Masses are the structures, the objects that have a three-dimensional form – usually vertical – and can be plants or solid structural features. Voids are the spaces not filled with masses and usually horizontal. Examples include pathways, lawn, areas of gravel or low ground cover. In other words, where you want 'stuff' to be and where you do not. It is probably the single most important thing to consider.

This balance can vary depending upon what you are aiming to achieve. 'Jungle'-style exotic gardens, for example, would have a far greater proportion of masses to voids. The sense of enclosure is important to the overall look and feel: the tension and anticipation of being led along a densely planted and enclosed pathway, then stumbling across a central 'clearing' – a void. This all adds interest and drama, although it is important not to lose sight of the scale and proportion of your design as it must sit happily in the space available. There is a danger that, by the end of summer, the planting in jungle-style gardens will have grown into amorphous blocks of foliage with pathways that are far too narrow to negotiate comfortably.

Schematic representation of masses and voids where masses are dominant.

Schematic representation of space where masses are less dominant.

Mike Clifford's Dorset garden is a great example of a mass-dominant jungle style design.

The Dry Garden at RHS Hyde Hall is a perfect example, keeping the masses lower to retain views over the outlying countryside and farmland.

For xeriscaping, however, the balance is tipped towards the voids. The nature of the plants and their growing requirements dictate that there should be space around them and between them. Planted too densely they become crowded, light-starved and etiolated, which, in turn, leads to weak plants that are not best placed to deal with the rigours of winter. Even simply thinking of xeric environments conjures up images of open, sunny spaces. With fewer masses more care needs to be used in arranging them as each one, and its relationship with the others, suddenly becomes more important.

Movement

While the proportion of the masses and voids gives the space balance, the way in which the masses and voids are arranged gives rhythm, or movement, through the space. You may want to make an introverted garden with a calm, central space. Conversely you may want to lead movement through the space towards a focal point or an as yet unseen part of the garden.

The design can be formal, with geometrical or strong shapes, often mirrored along axes within the space. An informal design would generally use more organic shapes, perhaps linking stronger shapes within the space. Formal designs would often have masses that are arranged symmetrically; conversely, informal designs would generally be more asymmetric in nature. Not always, however, as sometimes it adds more interest and surprise to have an asymmetric arrangement of masses within a formal space.

Schematic diagram showing a formal, symmetrical introverted space.

A section of the Sand Garden at York Gate, a beautifully designed introverted amphitheatre-type space.

Schematic representation of masses with an opening at the far end inviting movement.

My upper rock garden shortly after construction, almost an introverted space but with steps in the far corner leading the eye.

The same space, 16 years later, showing how the masses have increased but the sense of movement is still retained.

Schematic representation of an asymmetric and informal design encouraging movement and leading to a focal point.

Simplicity

It is important to give emphasis or interest but, at the same time, designs should be as disciplined as possible. Sweeping curves, straight lines and strong shapes usually work much

A client's garden shortly after construction. Raised beds allowed growing medium for both xeric and jungle-style planting to be accommodated.

A beautiful and harmonious composition at East Ruston Old Vicarage; the Desert Wash garden is an informal space filled with movement.

The design for the same client's garden incorporating several different features, such as a pond, vegetable garden and greenhouse, yet still keeping the overall space as simple as possible.

better than fussy, wiggly lines and sharp angles. Keeping non-essential features to a minimum will reduce the necessity for weak or awkward solutions for problems that should not really exist. You might fancy the idea of a pergola, bird bath, sundial, water feature and a lamp post or two, but trying to rationalise their places in the overall design might be a struggle.

These elements combine to give unity and harmony to a design. Only once this has been established should you then move on to consider the planting.

Planting Design

Designing the planting scheme for your garden involves many of the same processes employed when designing the whole garden but applied on a smaller scale. The balance of masses to voids, big plants to small plants must be considered. But other things come into play: pattern, texture, variety/repetition, contrast, emphasis are among other elements to be taken into account.

Masses and Voids Again

Within the context of the design, masses are represented by planting areas. These masses can, themselves, be reduced further into masses and voids – not just a solid block but structured taller, shorter, broader elements combining within the space of the mass. This adds interest and texture to the mass. Contrasting texture is important to give variety. For example *Opuntia sheerii* and *Aloiampelos striatula* will for some years be similarly shaped and sized masses, but they have completely different structural form and leaf pattern and, therefore, have different textures. So, broadly, they perform the same job within the context of the overall design but lead to contrast in the planting design.

Large, rounded pads of *Opuntia sheerii* give a broad texture to the mass it occupies.

A much finer texture is lent to the mass of *Aloiampelos striatula*.

Repetition

An example of repetition would be using the same plant, say *Dasylirion acrotrichum*, in different parts of the design so that your eye picks up the same mass and texture, which then leads to harmony. If you wanted to extend this repetition on a larger

Dasylirion acrotrichum makes a strong focal point, especially when repeated.

Pennisetum macrourum has a similar form and, therefore, useful when repeating the shape while also adding interest.

scale, however, using exactly the same plant might risk the scheme becoming monotonous. In which case you could consider pairing the dasylirion with a stiff-leafed grass, such as *Pennisetum macrourum* or maybe a young *Yucca rostrata*. For the first few years from planting these will occupy the same kind of space in terms of their mass and have similar textures, in that they will be domes comprising small, thin leaves. They are different, however, and, therefore, will achieve the goal of repeating a strong element in the design without the risk of introducing monotony.

The planting within the walled garden at Cotswold Wildlife Park, near the meerkat enclosure, has a repeating theme of similar, but different, columnar forms.

Colour

Broadly speaking colour is usually one of the key tools of planting design as it can be used to manipulate and excite the mind in so many ways. For xeriscaping, however, the palette of plants at your disposal is of necessity narrowed, so opportunities to design colour-themed plantings are much more limited and difficult to achieve. Succulent plants – at least, the hardier ones – tend to be green or glaucous grey with occasional purple and reddish tints. Their flowers, while often extremely colourful, I consider to be a much welcomed but passing bonus, so I tend to rely on contrasts of form and texture with colour provided by companion planting such as bulbs, e.g. *Tulipa linifolia*, or annuals, e.g. *Eschscholzia californica*. There are notable exceptions, of course, with cultivars of such plants as *Lampranthus spectabilis* being available in a range of hues so that broad sweeps of vibrant colour can be possible.

A blaze of colour throughout May and June from poppies and mesembs.

Hard Landscaping

Much of the hard landscaping areas of your garden contribute towards providing the voids in the design. Pathways, patios and areas of scree all help to connect and contrast with the masses of the planting and other structures. Again, material choices will depend on many factors – compatibility with the design and budget being foremost.

Patios

Natural stone is, of course, the premium material for patios and seating areas with both rustic and modern-style slabs available that are suitable for most designs. Purchase and installation costs are high but properly laid, a natural stone patio will last for decades, developing a patina over the years that is hard to beat or mimic. If you are not planning to move house it is always an investment worth considering. At the high end of the market, textured concrete slabs are often difficult to tell from the real thing and are usually cheaper, although recent years have seen some stone patio slabs, such as Indian sandstone, become more competitively priced. Timber decking is a quick, cheap and effective way to make a patio, especially on uneven ground or gradients, but even treated timber has a limited usable lifespan of around ten years at best. Modern composite decking materials have come on in leaps and bounds in recent years, the top-end

Anything other than natural stone paving in the garden of this Georgian terrace would look out of keeping.

A raised deck of composite boards in my garden makes a good facsimile of timber but should last decades.

products often appearing almost indistinguishable from the real thing, with complex and varied grain patterns pressed into the surface. They are non-slip when wet, easily cleaned and have an almost indefinite lifespan, but are expensive to buy and instal.

Paths

Moving from one part of the garden to another requires a pathway of some sort. Within the constraints of your design the choice of material will often depend on the frequency and heaviness of use, but with due consideration given to the shape of the path. Main thoroughfares for footfall and movement of materials using barrows, etc. are best constructed from solid materials. Stone or concrete slabs are perfect for straight paths but curves and more organic shapes can prove awkward when cutting slabs to size with associated huge labour costs. In these cases, paviours may provide a practical and flexible solution. They are available in many materials from granite setts and clay bricks to concrete driveway blocks.

Aggregate materials are the most flexible in terms of filling the path shape but not so easy to walk on or for wheeling

Concrete paviours offer a flexible solution for durable but irregularly shaped pathways.

High-grade plastic grids, designed for use with off-street parking, retain aggregate and make it easier for foot traffic.

machinery. Some self-binding materials, such as hoggin and MOT-type 1 sub-base, can be compacted to a hard surface but are perhaps rather utilitarian in appearance if not used carefully. A compromise can often be reached by using paving slabs as stepping stones within an aggregate pathway. For a more modern feel, materials such as steel or glass-fibre

grating, used in the building of raised walkways in many industrial and agricultural situations, can look extremely effective. Despite being heavy-duty, they appear light because of the open nature of their construction. Although massively popular for jungle-style exotic gardens, I think bark mulch paths are best avoided, they just do not sit comfortably with the overall dry climate look.

Mulch

Mulching around plants in your borders is not only aesthetically pleasing for a xeric garden but also has some important functions. Aggregates from natural sources are usually the best material to use. Stone and gravel, anything from sharp sand through all-in ballast to 40mm gravel, river-washed or beach cobbles, paddle stones – all work well visually. They help to reduce the surface germination of weeds. They perform an important role in water conservation, preventing water from evaporating from the soil surface. Just as importantly they allow free passage of rainwater, which just runs down the sides of the material and into the soil. By using a mix of different-sized or -shaped materials it is possible to mimic a natural scree hillside or river-washed *arroyo* quite effectively, both of which are eminently suitable for xeric planting. Organic mulches such as chipped or shredded bark should be avoided as they hold on to a fair proportion of the water and the damp material then just sits around the base of the plants. Also, as with pathways, it just does not really look right with the xeric plants.

ABOVE: Steel grating makes a durable, effective and strikingly unusual alternative for pathways, especially over uneven ground.

RIGHT: Stone mulch is perfect both aesthetically and functionally for the xeric garden here at East Ruston Old Vicarage.

Props

Introducing a little whimsy into the garden is never a bad thing but, of course, it is so very subjective. One way to do this is by the use of props. A few years ago there was an outbreak of longhorn cow skulls on the market, probably the most suitable of all props I have seen for xeriscaping. In my own garden, I am blessed with badgers and, faced with my newly purchased skull, they simply regarded it as a food source and chewed it to bits. There are still one or two around, though, which make me grin every time I see them. I am a fan of rusty metal; to me random objects left discarded conjure visions of desert towns in the south-western states of America. Not everyone agrees; some artist friends of mine are mortified by my lack of taste. Another opportunity arose when I accidentally allowed a patch of rot to expand in the side of my tallest columnar cactus, leaving quite a large and unsightly hole. A friend suggested that an owl would look perfectly at home there as, in nature, owls will often utilise such holes as nesting sites. As luck would have it, the next boot sale I went to provided the perfect thing.

RIGHT: A longhorn cow skull on a sun-baked border at Crete Lodge.

Perfectly at home in an otherwise unsightly scar on this cactus.

It is possible to go a little too far sometimes.

Use of Containers

This is a massive subject; one to which I cannot possibly do justice in the context of this book. Cacti and succulents generally have strong forms and present almost endless opportunities and challenges for finding suitably shaped containers in which to grow them to accentuate their particular characters. I think it is the most important aspect of container-growing succulents – not grabbing the first pot that comes to hand but matching pot to plant.

A cylindrical pot is perfect when topped by a tuft of dasylirion, with the caveat that stability needs to be considered when raising larger, heavier plants in taller pots. There has to be some ballast at the bottom to counterbalance the arrangement.

Wide, shallow pans are perfect for small, clustering plants such as *Aristaloe aristata*, as they are unlikely to tip over despite the small, flat footprint of the base of the pot.

The spots and speckles of both pot and plant are beautifully matched in Chris Grundy's garden.

A tall, slim pot is just the perfect vehicle for showing off the strongly recurved habit of *Mangave* 'Bad Hair Day' in Chris Grundy's garden.

Steve Buckley perfectly matches the spiral habit of *Aloe polyphylla* with the swirl of this ceramic.

Of course, growing plants in pots means they are mobile, even if, due to the size of the pot, it means using a sack barrow or even a pallet truck to move it. This allows more tender subjects to be grown outside during the frost-free months and opens the door to a whole new world of shapes and sizes. All you need is sufficient space, with appropriate ease of access, to house the plants over winter. Many interesting plants are content with being kept in 'just' frost-free conditions or even under unheated glass, as long as they are dry.

Whatever your choice of plant or pot, the growing medium has to be considered and, just as when growing in a raised bed, drainage is the watchword. The idea of using 'crocks' at the bottom of the pot has now been largely discredited but it certainly makes sense to still cover the drainage hole of the container to prevent the growing medium from falling through – at least until the plant is fully rooted. A regular application of liquid feed during the warmer months will keep your plant healthy.

A framed tapestry of echeverias displayed on an easel at RHS Wisley.

Gardens on a big scale require equally grand solutions. Here at RHS Hyde Hall this giant pot meets its match with *Aeonium* 'Zwartkop', artfully combined with *Dichondra* 'Silver Falls'.

Air plants tied to chicken wire fastened to a frame give a delicate effect.

ABOVE: A tasteful assemblage of pots at York Gate.

RIGHT: Here on St Michael's Mount careful steps have been taken to ensure a good display of pots.

BELOW: Steve Buckley, again, is a master of eccentric pot combinations and displays.

CHAPTER 4

XERISCAPING IN ACTION

Here I show a few examples of xeriscaping from selected gardens across the UK, ranging from small corners to complete landscapes. I decided early on not to include what is generally considered to be the jewel in the UK's exotic gardening crown: the Abbey Gardens on Tresco. The climate there is so benign, and the range of plants they can grow there is so huge, that it is almost the equal of the famous Italian and French Riviera Gardens. I believe there is more value in showing an agave winning its battle against the elements in Yorkshire than enjoying an easy life in the Scilly Isles.

Gardens Open to the Public

St Michael's Mount

The gardens wrap around a twelfth-century castle built on top of a huge granite outcrop sitting in Mount's Bay, a short walk along a tidal causeway from the town of Marazion in southwest Cornwall. This microclimate is exploited to the full, with many tender succulent plants thoughtfully planted, often appearing to sprout directly out of the immense granite boulders.

ABOVE: St Michael's Mount: the colour combinations here are stunning.

LEFT: St Michael's Mount: every nook and cranny is filled with a plant of some sort.

OPPOSITE PAGE: A sumptuous selection of succulents on St Michael's Mount.

The Minack Theatre

An open-air theatre carved into the side of a rocky cliff on the extreme south-west tip of Cornwall, this unique garden weaves around the terraced seating in what has to be one of the most dramatic settings of any garden. Frosts are rare, allowing a special blend of succulent planting that revels in the mild, wet and windy conditions.

The Minack Theatre: gloriously bright and cheerful planting.

The Minack Theatre: *Agave salmiana* var. *ferox* set in a sea of colour.

Ventnor Botanic Garden

A short ferry ride from the mainland, Ventnor Botanic Garden is set within the narrow undercliff strip on the south edge of the Isle of Wight. Summer temperatures there are high, winter temperatures extremely moderate and, with low rainfall levels, the climate must represent one of the very best locations for growing cacti and succulents outside in the UK.

BELOW: Ventnor Botanic Gardens: a hot, dry slope filled with succulents and other exotics.

Ventnor Botanic Gardens: puyas and palms make perfect partners.

East Ruston Old Vicarage Garden

Situated on the north Norfolk coast in a small pocket of benign microclimate, this 32-acre garden is as varied as it is vibrant. Of special interest here is the Desert Wash; a skilfully landscaped acre of graded flints, housing one of the finest outdoor collections of established yuccas, nolinas, dasylirions, agaves and puyas in the country.

East Ruston Old Vicarage: a masterclass on designing with succulent plants.

East Ruston Old Vicarage: growing on the edge for 25 years.

Tremenheere Sculpture Gardens, near Penzance

World-class art installations aside, this is one of the most exciting and sheltered gardens on the UK mainland with an unparalleled range of rare and exotic plants, many of which reflect the travels of the garden's founder, Neil Armstrong. The sunny upper slopes feature an impressive range of succulent plants enjoying the big skies and mild climate.

Tremenheere Sculpture Gardens: so many unusual plants thriving in the balmy Penzance climate.

Tremenheere Sculpture Gardens: pulse-racing planting with a view to die for.

The World Garden, Lullingstone Castle

Born out of the despair of kidnap and enforced captivity, Tom Hart Dyke's masterful garden boasts an astonishing number of plants from all around the world grouped geographically. Of principal interest is 'Mexico': a large, raised border featuring a selection of cold-tolerant succulent plants, made possible by a large winter cover.

The World Garden, Lullingstone: the Mexican border in all its glory.

The World Garden, Lullingstone: spiky plant heaven.

York Gate Garden

An inspirational garden just outside Leeds, effectively divided into a series of small 'garden rooms' giving visitors many varied experiences. The relatively new sunken 'Sand Garden' is a beautifully designed terraced amphitheatre with a range of unusual drought-tolerant plants growing in rubble and sharp sand to mitigate the high Yorkshire rainfall.

LEFT: York Gate: one section from the ground-breaking Sand Garden.

RIGHT: RHS Hyde Hall: established *Aloiampelos striatula* providing floral impact in the Dry Garden.

York Gate: lovely display of potted succulents.

RHS Hyde Hall

My local RHS garden, with which I have strong ties, features the iconic 'Dry Garden', which, since its inception in the late 1990s, has inspired gardeners throughout the dry south-east of England. Following recent redevelopment of the hilltop buildings, opportunities arose to push the boundaries with the construction of a new bed featuring some exotic xeric plants, with plans to add more succulent plants in future years.

RHS Hyde Hall: the exciting new xeric plant border at the hilltop.

Private Gardens

Melissa Scott

Melissa's garden, Crete Lodge, is an inspirational and beautifully curated array of specimen succulent plants, all growing happily in her freely draining flinty Norfolk soil.

Melissa Scott: a magnificent display of impeccably planted spikies at Crete Lodge.

Steve Buckley

Multi-talented sculptor, gardener and designer, Steve has a gift for matching the perfect pot to each plant, often to quirky effect, and then arranging the pots for maximum impact in his renowned London garden.

Alison Kent

Plantsmanship of the highest order together with an artist's eye is a winning combination. Ali moved established specimens from her previous garden, reconfiguring them to perfect effect in her new north Essex home.

Alison Kent: new xeric border planted with the most fabulous, mature specimens.

Steve Buckley: eclectic pots, xeric planting and a pathway enticing you in further.

Bill Howarth: a perfect island bed with one of the nicest *Yucca linearifolia* you could wish to see.

Bill Howarth

Several areas of Bill's remarkable Buckinghamshire garden have been given over to planting yuccas and other drought-tolerant exotics but this island bed in his front garden is outstanding.

Chris Grundy

As well as a choice selection of beautifully grown outdoor succulents, planted within a creatively designed garden in south London, Chris also holds the UK's premier collection of mangaves.

Chris Grundy: a sweeping curve featuring sumptuous succulents planted in a funky gabion.

Wayne Meyers

A superbly conceived and planted raised bed in Wayne's west London garden with some choice and well-grown agave specimens, and featuring excellent use of corten steel accessories.

Wayne Meyers: a raised bed perfectly set with some chunky agaves.

Maya Nedeva

A thoughtfully constructed area of raised beds in Maya's North Wales garden contrasts the look of a traditional rockery with modern gabions to striking effect, perfectly combining form and function.

Maya Nedeva: a brilliant modern twist on using raised beds in the wet North Wales climate.

Joel Robinson

A seasoned devotee of xeriscaping in the Pennines, Joel's new garden comprises a series of south-facing terraces that have been built into the steep slope that rises up from the house, presenting exciting opportunities for cactus and succulent planting.

Joel Robinson: one of many terraces, each with a fabulous selection of plants – especially that *Soehrensia*!

Kristofer Swaine

YouTuber Kris grows to perfection an astounding array of palms and other exotic plants in his Wakefield garden, among the most surprising of which are some beautiful yuccas and agaves.

Kris Swain: a cheerful corner of succulents beyond the jungle on a rainy day.

Richard Darlowe

Veteran exotic gardener Richard has been pushing the boundaries of what can be grown in the Barnsley climate for three decades with his Mediterranean-style planting, thoughtfully combining palms, yuccas and agaves.

Richard Darlowe: a little piece of the Mediterranean in Yorkshire.

Ed Croft

Ed's established garden in East Sussex contains a mind-boggling array of mature palms and other exotics packed in together with a wide selection of agaves.

Ed Croft: so many fantastic plants!

Rob Stacewicz

A fabulously creative plantsman's paradise; designer Rob put together some sensational combinations of the choicest plants within a spectacularly effective design, highlighted by many retro details spilling over from his mid-century-inspired Surrey home.

David Matzdorf

A groundbreaking green roof that features a naturalistic blend of drought-tolerant shrubs, perennials and cold-hardy succulents to breathtaking effect, recreating the feel of a Mediterranean hill-side on top of his unique house in north London.

David Matzdorf: cutting edge creativity and innovative green roof planting.

Rob Stacewicz: garden design and exotic planting of the very highest standard.

Darren Hobbs

A steeply sloping, south-facing site within view of the Bristol Channel, Darren's Newport garden is home to a bewildering array of the most unusual and unexpected plants, some of which are given careful winter protection where necessary.

James Knock

A long, flat garden near Brentwood in Essex, cold in winter but hot in summer, has been given interest and structure using gabions for raised beds. This has enabled a surprising number of succulent and exotic plants to thrive.

LEFT: Darren Hobbs: this large *Macrozamia communis* is just one of Darren's many champion plants.

BELOW: James Knock: a fine selection of palms and succulents grown hard in a raised gabion border.

PART II

THE PLANTS

The number of succulent plants that it is possible to grow outside in the UK is actually quite mind-boggling. A scant handful has been grown historically but largely it is without precedent. Over the past 30 years I believe I have tried a broader range of cacti and succulent plants outside than anyone, yet I still feel as though I have only scratched the surface. It really is an exciting frontier of horticulture, with so many species still to be tested in the context of a garden.

Within the scope of this book I cannot possibly cover all the plants I would like to include: I have left out so many as there simply is not the space. I feel an apology is particularly due to the smaller cacti: genera such as *Escobaria*, *Coryphantha*, *Austrocactus*, *Pediocactus* and *Sclerocactus*. There are literally scores of species that I have not yet tried, let alone those simply left discarded on the cutting-room floor. When trying to create an entire xeriscape I have, of necessity, used plants with more presence that are big enough not to be overlooked. On a smaller scale, these little cacti would add detail to a raised bed or container on a patio and can equally justify their place: it is just not what I have set out to achieve in my garden.

Of the plants I have tried, therefore, it is my aim to mention my personal favourites. I have highlighted those I consider to be the most suitable, having been tried and tested in my own and others' gardens, by calling them 'Keystone Species' and have given a more detailed description. Plants that did not quite make it to that list I discuss in smaller paragraphs under 'Worth Considering'. I have also added a category of 'Watch List' that more briefly mentions plants that do not have the same track record of being grown outside but show potential.

BELOW: *Denmoza rhodacantha*, a fine-spined form.

CHAPTER 5

CACTI

It has often been said that taxonomy is not so much a science as a war zone and, in my opinion, there is no bloodier battlefield than the naming of cacti. For example, it was put forward some decades ago that all the species previously classified in commonly grown genera such as *Trichocereus*, *Helianthocereus*, *Chamaecereus*, *Lobivia* and *Soehrensia* should be included in the genus *Echinopsis*, which, as a result, has become huge and unwieldy. This has been accepted by many authorities and rejected by others, leading to enormous confusion over the names of the plants. There may be valid science behind this but, for the gardener, important distinctions are being lost along with the old names. If you see a cactus in the genus *Trichocereus* the chances are that the plant will be tall and columnar. *Soehrensia* will be a small barrel and, in all probability, a hardy garden plant. By lumping these together you get absolutely no clues what a plant named *Echinopsis* could be like, the name now encompassing everything from the tiny peanut cactus to a towering giant. So, making no apologies, I just mention this to explain why I often use 'old-fashioned' names in preference, giving synonyms in the text.

Cacti are native almost exclusively to the 'New World', being found all the way from Canada to the tip of Patagonia and in a wide range of habitats. Some are among the most drought-tolerant of all plants, adapted to grow in desert conditions, whereas a number are much more temperate in their needs. Some are splashed by sea spray, others are found way above the treeline on some of the world's highest peaks. A few grow as epiphytes in moist jungle and forest environments. They have been introduced to many seasonally dry parts of the world, both as ornamentals and stock fencing but also, in the case of several *Opuntia* species, to use in cochineal production – described in more detail shortly. Sadly, in most countries where they have been introduced, opuntias have invariably slipped the leash, escaped into the wild and become a major ecological problem, outcompeting native plants and occupying vast tracts of land.

Feral opuntias in the Canary Islands have become a major ecological issue.

Broadly, all cacti are succulents, although historically the distinction is made from 'other succulents'. They are all in the Family Cactaceae; genera containing keystone species are given Subfamily and Tribe.

Areoles

A unifying feature of cacti, by which they differ from all other types of plants, is the presence of areoles. These are generally round, furry-looking bumps on the stems from which protrude the clusters of spines and hairs (and, in the case of opuntias, glochids), and are essentially modified and reduced shoots containing axillary growth buds. Flowers emerge from areoles, as do branches produced by shrubby cacti and leaves in those cacti that possess them. Place an opuntia pad flat on to the surface of some compost and it is from the areoles that the roots are formed.

Areoles on this cactus are very prominent.

Acanthocalycium

A small genus of cacti from Argentina, one or two of which have potential for the garden as they are found in fairly temperate regions. *Acanthocalycium violaceum* (syn. *A. spiniflorum* f. *violaceum*) has the longest track record for me and is worth considering. It grows into a small, half-grapefruit-sized plant showing much variation in the length and quality of spines. Flowers are generally pale lilac and it is reportedly hardy to –10°C. (H4)

Aporocactus

This is a small genus of epiphytic cacti of which one, from Hidalgo in central Mexico, has proved to be implausibly hardy. *Aporocactus flagelliformis* is a trailing plant, branching from the base, with many pendant, rope-like, finger-thick stems covered in fine bristly spines. In spring, clusters of showy magenta flowers burst from the sides of the stems. They are particularly suited to growing in hanging baskets, which is how a friend of mine, who lives in Ware in Hertfordshire, has grown a plant for 45 years. Around 2010, Mike decided that the plant had become too large to accommodate inside, as had been his habit, so he abandoned it outside that winter and left it to its fate. To his surprise it was still alive and unmarked in spring and has remained so ever since. In this time it has taken some very low temperatures, possibly as low as –8°C, and been covered with snow and ice. (H4)

Aporocactus flagelliformis in the Ware snow.

Chamaecereus

For many years just one species, *Chamaecereus sylvestrii*, was known, originally collected from north-west Argentina but not found in the wild since. *Chamaecereus sylvestrii* (syn. *Echinopsis chamaecereus*, *Cereus sylvestrii*, *Lobivia sylvestrii*) is the familiar 'peanut cactus' growing into a cluster of endearingly cute little finger-sized stems bearing, in early summer, so many vermillion flowers that it is almost impossible to see the actual plant. I would take a guess that nearly all cactus enthusiasts have had a crack at growing this at some point due to its forgiving nature and reliable flowering. Said to be hardy down to –10°C, it has survived –8°C in my garden when dry, although I have lost it during mild but damp winters if unprotected. (H5)

Cleistocactus

A genus of several species from Bolivia, one of which has been tried outside a few times in the UK. *Cleistocactus strausii* is a slim, columnar cactus, reaching perhaps 90cm tall and 8cm in diameter, branching at the base. It is covered in dense, short, white spines, giving the appearance of being furry and giving rise to its common name of 'silver torch cactus'. Flowers, if produced, are strange red tubes emerging from the sides of the upper portions of the stem. It comes from high altitudes, around 3,000m, and is said to be able to tolerate temperatures down to –10°C, if dry. My own plant is situated at the base of a south-facing wall and has taken –6°C. (H4?)

Cleistocactus strausii at the base of my house wall.

Cylindropuntia

Subfamily: Opuntioideae
Tribe: Cylindropuntieae

Once a part of the genus *Opuntia*, *Cylindropuntia* comprises between 20 and 35 species spread throughout the southern United States, Mexico and into the West Indies. Like *Opuntia*, they have jointed stems but, unlike *Opuntia*, the joints are cylindrical, not flat. They are branched shrubs, varying considerably in size from around 1m tall to 4m or more, infamous for their dangerously barbed spines that hook into anything. Collectively they are known by the indigenous name of cholla. Many of them are extremely cold-hardy but mainly intolerant of winter wet.

Keystone Species

Cylindropuntia imbricata

Common name: cane cholla, walking-stick cholla
Distribution: widespread through the southern United States and northern Mexico
Hardiness: H6/7

Cylindropuntia imbricata (syn. *Opuntia imbricata*, *Grusonia imbricata*) can be found from San Louis Potosí in Mexico to as far north as Kansas in the United States and is a conspicuous feature of the dry, low scrub areas it inhabits due to being evergreen and tall, reaching exceptionally to 4.5m high and across. The stems resemble plaited ropes that are easily detached at narrowing 'fault lines' at the joints between segments. The plentiful

Cylindropuntia imbricata.

spines are like needles with barbed tips that, like so many chollas, anchor themselves to the skin, fur or clothing of a careless passer-by. The brittle stem segments are then easily knocked off and transported away to form a new plant once released from the hapless carrier. Flowers are borne in early spring to summer and are vivid magenta.

I have grown this in my garden for twenty years and it has never been adversely affected by heat, drought, snow, rain or cold. It does, however, seem reluctant to hold its own weight, requiring the support of an iron stake to keep it upright. I suspect this is more to do with the surrounding plants growing up around it in the formative years and keeping the base of the cholla in too much shade: new plantings elsewhere in the garden are more promisingly erect. It is one of the hardier cacti, certainly one of the largest hardier cacti, with a very distinctive presence. It has a strong architectural shape, strong spine colour and gorgeously vivid flowers. It is also one of the more dangerous denizens of the succulent garden and, as such, needs to be sited carefully away from high-traffic areas.

In nature these will take down to a staggering –25°C or lower. Certainly my plant has survived my worst winters with absolutely no ill effects.

Worth Considering

Cylindropuntia echinocarpa

This is a stunningly beautiful cholla from the south-western USA and north-western Mexico growing to 1.5m or taller. It branches profusely, stems are covered in brilliant golden spines that sparkle in sunlight and invite stroking. That would be a mistake, of course, as it is every bit as dangerous as the rest. I grew it here for several years, losing it for no accountable reason. I aim to replace it. More sensitive to moisture than some others and less hardy, perhaps down to –10°C if dry. (H5)

Cylindropuntia echinocarpa.

Cylindropuntia leptocaulis

This is possibly the closest thing to barbed wire it is possible to grow. Pencil-thin stems clothed in vicious barbed spines in a tangled mess to 1m high. Yellowish-to-pink flowers can lead to red berries in winter,

Cylindropuntia leptocaulis in north-east Mexico.

hence the common name of Christmas cholla. I grew it here for a few years, during which time it experienced snow, rain and low temperatures of around –6°C, but I lost it during one exceptionally wet summer. Hardy to perhaps –10°C. I aim to replace it. (H4)

Cylindropuntia subulata (syn. *Austrocylindropuntia subulata*)

This is a robust, shrubby plant from the Peruvian Andes, often at high elevations, that presents itself as a bright green tangle of plump, bumpy spiny stems to head height or more if frost-free. It is unusual in that there are vestigial leaves that remain on the plant, as well as the yellow spines. In warm climates the stems are topped by small, pink-red flowers. It is hardy perhaps down to –5°C so possibly never a long-term prospect but, as it grows quickly, is worth considering for as long as it lasts and is easily replaced from cuttings. (H3/4)

Cylindropuntia subulata in Gran Canaria.

Cylindropuntia whipplei

This cholla is widespread throughout the southern states of the USA and up to Colorado. As a young plant this is similar to, but not quite as striking as, *Cylindropuntia echinocarpa*, with silver spines and a scruffier appearance. At maturity it becomes variable in size and amount of branching, as expected from a plant with a wide distribution. It is also more cold-hardy, said to take down to

Cylindropuntia whipplei.

–12°C or lower in habitat, and perhaps more tolerant of moisture. I grew one here for a few years but it did not ever gain momentum. (H4/5)

Denmoza

Subfamily: Cactoideae
Tribe: Trichocereeae

A genus of just one species from north-western Argentina.

Keystone Species

Denmoza rhodacantha

Common name: none applied
Distribution: widespread through north-western Argentina
Hardiness: H4/5

Denmoza rhodacantha (syn. *Cereus rhodacanthus*, *Cleistocactus rhodacanthus*, *Echinocactus rhodacanthus*, *Echinopsis rhodacantha*) is widespread in north-western Argentina, sometimes at high altitudes and often seen growing in quite temperate environments. As you would expect from a plant with a wide distribution, it is extremely variable in appearance, which explains, in part, the number of synonyms. Sometimes it is a short, squat globe, sometimes a thin barrel, sometimes a fat column with ancient specimens in the wild reaching up to 3m tall. Spines can be heavy, almost like those of a *Ferocactus*, or much thinner, also variable in colour from a russet-red (hence the name) to pale straw-yellow. The same plant can look different at different ages. Flowers, when produced, are odd red tubes almost identical to those of *Cleistocactus strausii* to which it is allied and with which it can hybridise in the wild. The name is derived from an anagram of the province of Mendoza, where it lives.

Denmoza rhodacantha is one of the more imposing cacti that can be grown in the UK. Together with the various *Soehrensia*, they eventually make stocky barrel plants with a form unlike anything else, large enough to make a real impact.

A robustly spined form of *Denmoza rhodacantha*.

The odd tubular flowers of *Denmoza rhodacantha*.

The spines are of huge ornamental value with red-spined variants, particularly, glowing intensely while still wet after a rain shower or back-lit by low evening or winter sunshine. It makes an important and impressive addition to the list of garden cacti.

In nature this will take down to below –10°C. In my own garden one survived my worst winter and had been growing outside for over ten years before dying for no apparent reason. I now have a few more.

Echinocactus

A genus of six species (but many more by some authorities) from the southern USA and north-eastern Mexico of which one is worth considering. *Echinocactus texensis* (syn. *Homalocephala texensis*) is a squat, shallowly domed, tough-textured cactus with extremely tough, curved, thick spines. It is known evocatively as the 'horse crippler' and found in the more arid parts of its range. It is easy to grow but very slow; as a result finding large enough plants to trial outside is both difficult and expensive. It is said to favour a highly drained growing medium in which it can take temperatures down to perhaps –18°C. In the UK it is largely untested but certainly one for which to watch out. I sadly wasted a good healthy specimen when, in my haste to experiment, I planted it in a totally inappropriate place and it rotted. (H5?)

Echinocereus

Subfamily: Cactoideae
Tribe: Phyllocacteae

A large genus of between 70 and 150 species, depending upon the authority, occurring in the southern states of the USA and across northern Mexico. They can be solitary or clustering, generally small individually but can grow into large groups. They are known collectively as 'hedgehog cacti'. Many are extremely hardy and make excellent garden plants.

Keystone Species

Echinocereus reichenbachii subsp. *baileyi*

Common name: lace cactus, hedgehog cactus
Distribution: Oklahoma and Texas
Hardiness: H7 in habitat

The species *Echinocereus reichenbachii* is numerous and widespread through southern and central USA and down to northern Mexico. Several varieties and subspecies have been recognised, of which, for my money, *Echinocereus reichenbachii* subsp. *baileyi* is the pick. It comes from the southern Great Plains and into the Chihuahuan Desert, growing in rock outcrops or even just in grassland.

I consider it to be the best small cactus it is possible to grow in the UK. Stems are around 4–5cm across and ultimately 15cm or so tall and plants can remain solitary or cluster into a small group. Spines are quite dense and usually creamy white. The flowers, produced in May/June, are quite astonishing: 7cm across and completely obscuring the plant when viewed from above. They are pale magenta with a dark throat and appear to have been made out of silk. Individually the flowers do not last very long but each plant can produce more than one and, if growing in a cluster, you can have a succession of flowers for a week or more.

In their habitat they have a dry, cold winter, taking temperatures down below –20°C. I have grown it in my garden and find it not the least bothered by –8°C and wet snow.

Echinocereus reichenbachii subsp. *baileyi*.

Echinocereus rigidissimus subsp. *rubispinus*

Common name: ruby cactus, rainbow cactus
Distribution: northern Mexico
Hardiness: H4/5

Part of a larger-ranging species, *Echinocereus rigidissimus* subsp. *rubispinus* (syn. *Echinocereus pectinatus* var. *rubispinus*) is arguably one of the most attractive forms as the extremely dense, very short, pink-red spines engulf the green body so thoroughly that the whole plant looks pink-red when seen from a distance. Stems are around 7cm thick, showing ridges or bands indicating growth cycles. Solitary at first, these stems can grow to 30cm or more in length and, with age, will often cluster into a small group. In flower it is spectacular, with enormous cerise to magenta, white-throated blooms up to 10cm across produced in early summer. In habitat it is found on steep, south-facing rocky slopes of neutral-to-acidic scree.

Said to be hardy to around –12°C, plants in my garden have seen –8°C without issue but, at the time of writing, have been dug up and destroyed by foxes for no apparent reason other than mischievousness. This is one that would benefit from a rain cover, if possible.

Echinocereus rigidissimus subsp. *rubispinus*.

Echinocereus triglochidiatus

Common name: king cup, mound hedgehog, Pitallito
Distribution: most of the southern USA and northern Mexico
Hardiness: H5/6

Echinocereus triglochidiatus.

This species has a massive distribution in the wild, which means that it can be extremely variable in appearance. Stems can be anything up to wrist thick, sometimes over 40cm tall and generally offsetting profusely to make a dense, multi-stemmed colony. Spines are generally straw-coloured but often age to black. The degree of spination is particularly diverse with a virtually spineless form, var. *inermis*, recognised. This is my first cactus to flower, usually mid to late spring. Flowers are cup-shaped and bright red – sometimes slightly washed out but more usually a startlingly intense orange-red. In habitat these are often found in mountain woodlands and areas with more shade than generally expected for a cactus. Nevertheless a position in full sun is best in the UK. My plant has survived my worst winter with no issue and no rain cover, and some forms are said to be hardy down to –20°C.

Worth Considering

Echinocereus coccineus

A small, clustering species that is widespread across the south-western USA and northern Mexico. It is ostensibly similar to *Echinocereus triglochidiatus* but generally a little smaller and spinier. It is less tolerant of winter moisture and best

Echinocereus coccineus.

grown with a cover, drainage and full sun being particularly important. A plant has taken –8°C in my garden but it is said to be able to tolerate far lower, especially if dry, perhaps down to –15°C. (H5/6)

Echinocereus viridiflorus

This species is the most northerly member of the genus, growing as far north as Wyoming and Dakota in a variety of habitats. It is variable in appearance but, typically, grows into a single, short column, to 25cm, or a small cluster of dumpy balls. Flowers are around 2cm across, produced from the lower half of the plant body and a pale, silky green. It is extremely hardy in habitat, taking down to –20°C; in my garden it has taken –8°C. (H5/6)

Echinocereus chloranthus is similar, having flowers that are a little smaller and inclined to be a dirty green.

Watch List

Echinocereus engelmanii is one of the most common species in the USA but I have not grown it for long. It forms colonies of densely spined little columns with bright pink flowers, often with a dark centre. *Echinocereus fendleri* is similar but more inclined to remain solitary. I have not grown *Echinocereus dasyacanthus* but intend to. It resembles *Echinocereus reichenbachii* in body but has yellow flowers. In habitat these are all incredibly hardy, down to below –15°C. In the UK I would expect them to take at least –10°C or lower, but they may well be moisture-sensitive.

Echinopsis

A large genus of, generally, small-bodied, solitary or clustering cacti distributed across many South American countries and the cause of some of the fiercest taxonomic conflict in botany. Several are worth growing, the most commonly encountered of which is likely to be *Echinopsis oxygona* a species from south Brazil, through Uruguay and down to northern Argentina. It grows into a small barrel to around 30cm tall and, maybe 12cm across, offsetting at the base. Impressive long-tubed white or pale mauve flowers, slightly scented, open at night and last one day. Said to take down to –10°C if dry, plants have survived –8°C unprotected in my garden. (H5)

Echinopsis huascha (syn. *Lobivia huascha*, *Trichocereus huascha*) is another well worth growing. Generally the plant is a yellowish-green, droopy, narrow column, branching at the base, with short, dense yellowish spines. Flowers are large and variable, yellow or red and up to 10cm across. I have been growing this in my garden in a position that is too shady for twenty years. It has never flowered and grows only slowly but is still there, having endured my worst winters intact. (H5)

Echinopsis oxygona cultivar.

Ferocactus

A genus of around 30 species distributed across the southern states of the USA and throughout Mexico. Young plants are globular, ageing to different adult forms: some are large, solitary barrels, some clustering barrels, some form immense domed colonies of densely clustered heads. Most have impressive and colourful spines with a ring of various coloured flowers produced at the top of the plant. Several species are extremely hardy to cold temperatures but they are generally very moisture-intolerant, typically growing in very hot and dry conditions. I have tried and killed many species outside but I am currently having success with *Ferocactus hamatacanthus* (syn. *Hamatocactus hamatacanthus*), which has survived down to –5°C but, more importantly, a week-long period of icy snow, freezing then thawing on top of the plant during my early cold winter. If it carries on, I look forward to seeing it grow into a 60cm barrel, covered in extremely long, thin, white spines, with large 7cm yellow flowers followed by edible fruit. It is native to north-western Mexico and south-western Texas. (H4/5?)

I have recently read hardiness reports for *Ferocactus glaucescens* that suggest it has survived –12°C unprotected in south-west Texas. Native to central Mexico, it is said to be rather more moisture-tolerant than most *Ferocactus* species, so will be the subject of a trial in my garden very soon. If all goes to plan, it will make a stunning addition to the cactus garden, growing into a plump 60cm china-blue globe with straw-yellow spines and glossy, pale yellow flowers. (H4?)

A final mention must go to *Ferocactus hystrix* as I have had a large, potted specimen of this growing outside in my covered but open-sided front porch for nearly twenty years. It does not really get enough direct sun but nevertheless flowers each year and it is now a whopping 40cm diameter globe of spines waiting to puncture an unwary postman. Kept dry over winter this plant has been through all my worst winters. It is so big, old and

Ferocactus glaucescens in my greenhouse.

Ferocactus hystrix.

nice that I have never felt inclined to plant it outside but would certainly do so with another, should I find one at a reasonable price. (H4/5?)

Gymnocalycium

A large genus, widespread throughout much of South America, a few of which grow in more temperate regions and are, therefore, worth considering. Of these I have experience of *Gymnocalycium gibbosum*, from Argentina, which is a small globular cactus with a dark green, almost black body, dark spines and off-white flowers. It was the first cactus I grew outside and it lived quite happily for over 25 years, taking my worst winter in its stride, only to unexpectedly succumb recently during an unprecedented long, hot, dry spell. (H4/5)

Gymnocalycium gibbosum.

Maihuenia

A genus of two strange species from extremely high altitudes in Chile and Argentia, being different enough from other cacti to be tucked away by themselves in their own subfamily. In appearance the two are very similar: individual stems are initially finger-sized, coarsely spined affairs, eventually clustering into immense tightly packed cushions. *Maihuenia poeppigii* is the most likely to be encountered, this is the smaller of the two and has yellow flowers. *Maihuenia patagonica* has white to mauve flowers. I mention them both as they are more tolerant of moisture and cold than practically any other cactus but, for me, only really become interesting when they are a couple of decades old and grown into a mat. Until then they look rather like scrawny abandoned baby hedgehogs. (H5)

Maihuenia poeppigii at RHS Wisley.

Opuntia

Subfamily: Opuntioideae
Tribe: Opuntieae

Opuntia is one of the largest genera of cacti. Authorities fail to agree but there are anything up to 300 species and many more natural hybrids, distributed across the New World but also naturalised all over the globe where climate allows. These are the flat-padded prickly pears, often called tuna, which, over the centuries, have provided fodder for livestock and humans with food from pads and fruit. I have eaten both in Mexico and became quite fond of them. The ripe fruit, when cleared of the tiny spines, appears to be something like a cross between a watermelon and a shotgun cartridge with dense, sweet flesh that is packed full of hard, round seeds. It is, however, a tasty and thirst-quenching snack when clambering over a dusty hill-

Opuntias bear the most exquisite flowers.

Nopales in a Mexican market.

side. The paddles – nopales – are again cleared of spines, steamed or blanched then griddled. You can often buy them in jars, pickled, where the mucilage that has come out from inside the pads leaves slimy strings as you attempt to take them out of the jar, which is revoltingly amusing to children and adults alike.

Typically they grow into branched shrubs, thickening at the base, composed of flattened pads (cladodes) stacked upon each other, branching as they go. One of the notable features of opuntias is the little dots of minute hooked spines, called glochids, that are clustered around the base of the spines in the areoles. Despite looking innocuous they need to be treated with the utmost caution! Once in the skin they are almost impossible to get out and can cause nasty and long-lasting irritation. Some opuntias are the most cold-hardy of all cacti, with one species, *Opuntia fragilis*, being found as far north as central Canada.

With such a huge number of species found over such a vast geographical range it comes as no surprise that there are many opuntias that can be grown successfully in temperate climates. It boils down to personal preference and, perhaps, availability.

Cochineal

Various *Opuntia* species are host to the cochineal bug – a tiny parasitic scale insect that is the unlikely source of a vivid, much sought after and, at one time, exceptionally valuable natural red dye that is often simply called cochineal. In its heyday this was produced by the tonne exclusively in the Mexican state of Oaxaca, the dried bodies of this humble little pest actually becoming Mexico's second most valuable export behind silver. The advent of cheap artificial colourings in the late nineteenth century all but wiped out the industry, although there has recently been an upturn in demand with increased interest in natural and artisanal products, plus concerns that artificial colourings might be carcinogenic.

Cochineal bugs on a Canarian opuntia.

Keystone Species

Opuntia scheerii

Common name: none applied
Distribution: uncertain, believed Durango, Mexico
Hardiness: H4/5

Opuntia sheerii is a small shrub, growing to around 1.2m high, comprising branched layers of oval pads that are grey-green in colour. These pads are densely covered in very short, thin, straw-coloured spines and also soft hairs. The glochids are so short as to be barely noticeable. Short-lived flowers are a pale lemon-yellow on day one, changing to peach on day two. It comes from Mexico, although exactly where is a matter of conjecture as the exact same plants as those in cultivation have not been found in the wild.

Opuntia sheerii.

I have had *Opuntia sheerii* in my garden for over fifteen years and it is, by some measure, my pick of those *Opuntia* species I have tried. It is one of the few cacti I grow that it is possible to stroke with my hand and not need minor surgery afterwards, the spines being flexible enough to bend slightly. It is best, however, not to pinch or grasp them as your skin will soon find the glochids. It is also the only *Opuntia* I have grown for any length of time that remains erect: usually they collapse when reaching around four or five storeys high. Which is such a shame as it is that unique outline that, for me, makes growing opuntias so appealing.

Close view of areoles showing fine spines and hairs.

I have found it to be much hardier than generally believed, taking my worst winters in its stride with little-to-no damage, even unprotected. Kept dry, Tom Hart Dyke reports his plant was completely undamaged at –16°C when under cover in his 'Mexico' border in January 2011.

Worth Considering

Opuntia basilaris

This is a variable species from across most of the southern USA, generally shorter than most opuntias and, therefore, not as inclined towards scruffiness. Pads are oval or heart-shaped, quite plump, blue-grey often with a tinge of purple, the surface dotted with glochids but with few or no spines at all. The common name of 'beavertail cactus' gives the overall impression nicely. Flowers are a vivid magenta and among the showiest in the genus. It is extremely hardy to cold in nature, seeing temperatures down to –20°C or lower, but moisture can be a problem. I lost an established plant one very wet winter some years ago in a less than ideal substrate and have, for some reason, only recently got around to replacing it. (H5/6)

A rather scruffy *Opuntia basilaris*, just starting to establish.

Opuntia cantabrigiensis (syn. *Opuntia engelmannii* var. *cuija*)

This is one of the better opuntias, keeping reasonably upright for some years and growing vigorously. For decades there has been a large clump of this growing outside the glasshouses at Cambridge Botanic Gardens, from where I begged a cutting twenty-odd years ago. Some authorities would have this as a variety of the widespread and variable species *Opuntia engelmannii* but I like to keep the clone separate using this name. In habitat these are very hardy, down to –20°C or lower. My plant took my worst winter with no issues but was irreparably damaged when I carelessly trampled over it when stepping down from a ladder. (H4/5)

Opuntia cantabrigensis.

Opuntia chlorotica

I have recently planted a rather handsome clone of this species. I am hoping it works out because it is exceptionally pretty with huge, round, blue-ish purple pads, attractive spines and, hopefully one day, large, bright yellow flowers. (H4?)

Opuntia chlorotica.

Opuntia fragilis

This grows naturally further north than any other cactus, found all the way up from the mid-western USA to as far as British Columbia. Plants are small, individual, pale-green pads the size of quail's eggs and easily broken off from the main plant. Flower colour is variable, usually yellow but can be fuchsia pink. If anyone doubted the hardiness credentials of this cute little cactus there is a lady in Finland,

Opuntia fragilis growing happily in Finland.

Pia Leppänen, who grows this species outside with no protection in her rock garden, the plants taking –30°C with a covering of snow. (H5/6)

Opuntia humifusa

This is distributed along the eastern side of the USA from Florida to Massachusetts in dry, sandy or rocky environments. It grows into a lax plant, the dull green pads barely holding themselves upright, and slowly sprawling to make a decent-sized clump to around 1.5m. The yellow flowers are its saving grace, freely produced in a sunny spot. I mention it as one of the most forgiving of opuntias, tolerant of extreme cold and wet, and hence often recommended but I do feel that, unless extreme hardiness is of paramount importance, there are more attractive species deserving of garden space. (H5/6)

Opuntia humifusa.

Opuntia macrocentra

This is a highly ornamental, low-growing species with round, glaucous, grey-green pads, tingeing purple when cold or drought-stressed. Long spines are present on the upper and outer parts of the segments. Flowers are generally yellow with a red throat and quite striking. Not as cold-hardy as most, said to tolerate down to –12°C if dry but less in a wet climate. I have never kept this alive for long but keep trying, the current plant having endured –5°C and a week-long freeze with no apparent issues.(H4)

Opuntia macrocentra in a nursery in Arizona.

Opuntia phaeacantha

Possibly the most common and widespread opuntia in the whole of continental North America. It usually slumps to a prostrate growth form, pads are quite thick and leathery, and the degree and heaviness of spination extremely variable. Flowers are generally yellow, sometimes with red throats or stripes. Most forms will be extremely hardy, taking down to –20°C or lower in habitat. I am sure some forms are excellent garden plants, I just need to search more thoroughly to find them. (H5)

Opuntia polyacantha

This is the plains' prickly pear with an enormous distribution in the wild. It is a low, spreading plant with flat, green pads up to 20cm long, often densely covered in spines and yellow or magenta flowers in early summer. In habitat it is hardy down to a mind-boggling –45°C in the northern Canadian reaches of its range. It is also said to withstand winter wet in perfectly drained soil. For reasons that I cannot explain, I have never grown it, but it is on my 'to do' list. (H5/6)

Oreocereus

A genus of eight to ten species, all from extremely high altitudes in the Andes, some from well over 4,000m, from Peru down to Argentina, characterised by clumping columns, often covered with dense hairs through which stout, golden spines protrude. In habitat they experience extreme cold but are somewhat susceptible to rotting in wet winters or high humidity. I have had some success with two species but currently only have one in the garden. At the base of a south-facing wall *Oreocereus celsianus* has grown into a large 80cm plant for me, the hair covering is somewhat wispy and scruffy and almost begging for a comb-over.

Oreocereus trollii, at least the ones I have tried, tend to be smaller, more inclined to clump, have much denser hair and are overall a much neater plant but, in my experience, a little more difficult to please. I lost my last attempt during my early cold winter.

Both can survive for several years but I have never kept one going into its second decade. I try to make a point of replacing those lost in the garden as and when I come across cheap replacements as I think their unique hairy appearance adds value and contrast to the planting. I have never seen flowers on either but, should they decide to flower one day, I would be treated to quite impressive reddish trumpets. There is a tendency for the white hairs to become quite dirty as the stem ages. I wonder how they would react to being shampooed one sunny day? Maybe this summer. (H4)

Oreocereus celsianus.

Selenicereus

Selenicereus is a genus of largely tropical epiphytic cacti with slender stems that snake their way into host trees and, usually, have enormous, nocturnal, showy flowers. These are often followed, in some species, by large, edible fruits known as dragon fruit or pitaya. Something of a curveball in the context of this book but there is actually a fairly hardy species, *Selenicereus spinulosus*, from eastern Mexico. It loosely resembles a braided cord 1.5cm thick with fine, bristly spines from which a large number of aerial roots emerge and root into whatever they can. My main plant, originally a cutting from pioneering plantsman Rob Senior over twenty years ago, lives full time inside my unheated greenhouse but, for the past decade or more, a small plant of it has been living in the crotch of a yew tree. It does not grow very quickly nor has it ever flowered but it is stubbornly alive and well, and unusual enough to be noticeable. I suspect a wetter climate would suit it more but it does OK. With the overhead cover from the yew branches it has taken down to –6°C. (H4)

Selenicereus spinulosus.

Soehrensia

Subfamily: Cactoideae
Tribe: Trichocereeae

A genus of two, possibly three, species from the high mountains of north-western Argentina, soehrensias are one of the very best cacti for growing out of doors in the UK. They are certainly among my favourites. They grow from small globes into imposing robust barrels, have beautifully coloured spines and flowers, and are tolerant of cold, wet, snow and ice.

Keystone Species

Soehrensia bruchii

Common name: none applied
Distribution: north-western Argentina up to 3,500m altitude
Hardiness: H4/5

Soehrensia bruchii (syn. *Echinopsis bruchii*, *Lobivia bruchii*, *Trichocereus bruchii*, *Lobivia formosa* var. *bruchii*) is an impressively squat barrel, around 30cm across, often clumping, sometimes remaining solitary (and reaching over 1m high), mid-green with prominent ribs and yellow spines. The top of the plant is depressed into a kind of bowl and it is usually within, or just outside of, this bowl that the flowers are borne in late spring to early summer. Flower colour is variable from bright pillar-box red to a burnt orange, usually 5cm across. It grows in nature in grasslands, scrub or rocky plains, with parts of its range being a reasonably temperate environment.

Soehrensia bruchii, burnt orange form.

I have grown *Soehrensia bruchii* in my garden for twenty years, adding more and more plants as its accommodating nature has unfolded. I have only ever lost one and I believe that was due to it being a weak plant rather than the fault of the species. Despite this I still sometimes find it hard to believe that it is possible to grow such a chunky and imposing cactus outside. Grown in full sun the spines develop a much stronger colour and the whole plant then becomes a passable imitation of the tender golden barrel cactus until it gives the game away with its vivid flowers.

My original plant was completely unfazed by my coldest winter. Subsequent plants have been fine through both my early and late cold winters too. I have no doubt it would take much lower temperatures if unprotected and certainly lower still if kept dry.

Soehrensia bruchii, bright red form.

Soehrensia formosa

Common name: none applied
Distribution: north-western Argentina into Bolivia
Hardiness: H4/5

Soehrensia formosa (syn. *Echinopsis formosa*, *Lobivia formosa*, *Trichocereus formosus*, *Acanthocalycium formosum*) is similar in shape and stature to *Soehrensia bruchii*, again sometimes solitary, sometimes clustering, but the spines are much denser and longer. These spines can be yellow, straw-coloured or pale enough to appear almost white, and in some individuals so densely covering the plant that you cannot actually see the body at all. Flowers are produced at the top of the plant, nestled within the spines, generally

Soehrensia formosa.

A group of *Soehrensia formosa*, spines glowing in low evening sunshine.

yellow within and burnt orange on the underside, but can vary from pinkish-yellow to red. I have read that ancient specimens can reach over 1.5m high. Their habitat is higher, colder and more xeric than that of *Soehrensia bruchii*, reaching over 4,000m altitude.

I absolutely adore this plant. It has everything I look for in a cactus: hardiness, attractive spines, brilliant flowers. I was fortunate enough to acquire an exceptionally ancient specimen from a nurseryman in the Netherlands who specialised in acquiring entire collections from those giving up their plants. It is 60cm tall, 30cm across and in all probability the oldest plant I own. I say this confidently as I have one documented to be 70 years old that is half the size. I pass it every day and every day it puts a huge smile on my face.

A more recent addition than my *Soehrensia bruchii*, it has taken both my early cold and late cold winters with no ill effects; I am certain it will take still lower temperatures. I have several others, here and there around the garden; maybe they, too, will grow to become like this. When backlit by late evening sun, the spines light up like a halo and I can think of no more stunning effect within the xeric garden than this.

Trichocereus

Subfamily: Cactoideae
Tribe: Trichocereeae

Many authorities subsume the genus *Trichocereus* into *Echinopsis* but I feel important distinctions are lost by doing this so, in the face of what some would call more modern thinking, I still cling on to the old name. *Trichocereus* are generally tall columnar cacti distributed throughout South America. A number of really interesting and potentially hardy species are found at high altitude in the Andes of Peru, Bolivia, Argentina and Chile, and it is without hyperbole that I say they provide us with the most stunningly unusual and dramatic plants it is possible to grow outdoors in the UK.

Keystone Species

Trichocereus candicans

Common name: none applied
Distribution: north-western Argentina to 2,000m altitude
Hardiness: H4/5

Trichocereus candicans (syn. *Echinopsis candicans*) is a small columnar cactus that has a leaning, spreading habit rather than

Trichocereus candicans.

growing erect and upright. Yellowish-green stems grow to 60cm long, 15cm or so across and branch freely at the base, forming a many-stemmed cluster. Spines are yellow-brown, variable in length and quite robust. Flowers are spectacular: white, scented, 20cm-long spidery trumpets, 15cm in diameter, with a pinkish-red back. Possibly one of the most impressive of any hardy cacti. In the wild it is found growing in open, high-canopy forest and scrubby grassland.

I have grown this in my garden for nearly twenty years, during which time it has endured my three bad winter events. My worst winter caused damage to the growth tip of one stem, which then promptly branched the following spring. It has slowly become a cluster of eight heads spreading to 70cm and flowers regularly. I have a plant of the same vintage, bought labelled as *Trichocereus pseudocandicans*, which is similar, a bit smaller and may or may not be the same. It is growing in too much shade and has yet to flower.

Trichocereus pasacana

Common name: Cardon
Distribution: north-western Argentina and into Bolivia up to 4,000m altitude
Hardiness: H4/5

Trichocereus pasacana (syn. *Trichocereus atacamensis*, *Echinopsis pasacana*, *Echinopsis atacamensis* subsp. *pasacana*, *Cereus pasacana*, *Helianthocereus pasacana*, *Leucostele atacamensis* subsp. *pasacana*) is a huge, columnar cactus: there are ancient specimens, said to be over 300 years old, reaching 10m high. They form quite dense forests on steep slopes at extremely high altitude. They are usually seen as solitary stems (they do branch but only when very old) that, on smaller garden-sized plants, are generally 25–30cm across. They have around twenty ribs and are, like many high-altitude cacti, more or less densely covered in thin, golden-yellow to straw-coloured spines. Flowers are pinkish-white.

Trichocereus pasacana.

Trichocereus pasacana: closer look at the top.

I have grown *Trichocereus pasacana* in my garden for something like eighteen years, during which period it has taken my worst three winters without difficulty. At the other extreme, because of the high montane provenance, these are said to resent temperatures above 30°C but my plant has recently been subjected to 40°C in an eight-day period of temperatures over 35°C, again without issue. It has not yet flowered but this is easily compensated for by the quality of its spines. They are incredibly variable in extent, length, colour and weight, which make them all the more fascinating: as a result I have, over the years, added several more plants of the species, as I find them particularly attractive, each with its own character. Full sun and a hard, rocky growing medium develops the best spines, even if this is at the expense of speed of growth. Nevertheless my largest plant can grow up to 20cm of stem in a season and it is interesting to note that spring and autumn spines are shorter than those made in summer, making it possible to see annual growth cycles in the stem.

I would say this would make my top ten garden plants short-list, and the competition is fierce!

Trichocereus pasacana: detail showing the gorgeous spines and areoles.

Trichocereus schickendantzii

Common name: none applied
Distribution: north-western Argentina up to 3,200m altitude
Hardiness: H4/5

Trichocereus schickendantzii (syn. *Echinopsis schickendantzii*) is a sprawling, slender, columnar cactus with stems reaching 80cm long and 8cm across. It branches profusely at the base and also along the stems where they touch the ground. Stems are mid-green with narrow ribs and covered in short, dense, yellow spines that age to brown. Flowers are produced near the top of the stems and are, in relation to the plant, quite enormous. On my plant they are pure white with a hint of pink on the reverse, 20cm in diameter and scented – opening at dusk and closing at dawn the next day. In habitat it grows in grassland, open forests and rocky slopes, and is said to take down to −10°C or lower.

Trichocereus schickendantzii flowers.

This is one of the most eye-catching plants in my garden. It has never shown any inclination to grow vertically; rather each stem seems to crawl along the ground but with the stem tip curving upwards. The branching is so numerous that stems mound over each other, somehow resembling the writhing snakes of a Medusa head. In twenty years the plant now covers an area of around 1.8 × 1.2m and shows no sign of slowing down.

I have never protected it; during my worst winter some of the stems were slightly damaged but it soon outgrew this and has since carried on regardless. One thing to be wary of is contact with the spines, which dislodge very easily and are next to impossible to remove, resulting in painful irritation that can last for several

Trichocereus schickendantzii.

weeks. I think if kept dry it would be easily hardy down to –10°C or lower.

Trichocereus terscheckii

Common name: Cardon Grande
Distribution: north-western Argentina and into Bolivia to 2,000m altitude
Hardiness: H4/5

Trichocereus terscheckii (syn. *Echinopsis terscheckii, Cereus terscheckii, Leucostele terscheckii*) is a giant of a plant. A single column to start with, plants can produce several branches – or arms – from the lower third of the body as they age, eventually reaching up to 12m high, while at the base the stem can be 40cm across. It grows in many of the same regions as *Trichocereus pasacana* but at lower altitudes where the climate is more temperate and less xeric: they are often seen together on the same slopes with a defined transition between the two. Stems are mid-green, fatly ribbed and fewer in number compared to *Trichocereus pasacana*. Spines are yellow, ageing to brown, variable in appearance but fewer and shorter than those of *Trichocereus pasacana* and overall covering the plant much more sparsely. Flowers are white, the size of teacups, produced along the sides of the stem in summer, opening late afternoon and lasting for about a day.

Trichocereus terscheckii, ribs.

Trichocereus terscheckii, flowering.

In 2003 I was fortunate enough to find for sale, at a very reasonable cost, a 2.5m-tall specimen. Since then it has grown considerably and taken all the weather that has come its way, including my worst three winters. This was without protection, of course, as even back then it was too tall to fashion any kind of practical covering for it. The growth rate I consider to be fast and seems to depend upon how wet the growing season is, but I would estimate the plant grows at around 15cm or so a year and now towers to an unimaginable 5m high. It flowers every year, one summer producing over 40. It is astonishing to think that something so tall and heavy can be stable in the ground: I used to stand and watch it on windy days, agonising over each sway of the stem, but now I do not even give it a second thought. To date it has remained solitary: every year I eagerly inspect the stem, looking for signs of impending arm development but nothing so far. It should be any time soon, I am sure, as I have seen plants in botanical gardens abroad smaller than mine that have already grown at least one.

It is my most famous plant, if plants can enjoy fame, having been featured in newspapers, magazines, books and even starred on television. It is simply the most spectacular and unusual plant to see growing in an English garden. Unwary visitors, not expecting to encounter a 5m high cactus, invariably just laugh – it being so completely outside of any other horticultural experience.

Worth Considering

Trichocereus chiloensis

This is a tall, clumping columnar from Chile that has slender stems and conspicuous bumpy ribs. Spines can be incredibly variable, sometimes as long as toothpicks. I have grown several but the best was a gift from a friend, Mark Morgan, who grew a few from seed he collected from a specimen at the very highest edge of its altitude range in Chile. His plants have been kept outside in pots for many years in the Midlands, seeing temperatures down to –6°C or lower. My one, a 1.6m-tall beauty that Mark somehow fitted into a tiny Toyota when he drove it down, has now settled in and is growing well, taking my early cold winter with no issues after a couple of mild ones. (H4/5)

Trichocereus pachanoi

This is a variable plant but generally a slim, columnar cactus – ribs are few but fat – that in time branches from the base. Some forms have sparse spines, others are completely nude. It comes from the Andes throughout much of South America, sometimes to 3,300m altitude. Said to be hardy to –12°C or lower, my plant at the base of a south wall has taken my worst winter with just a little marking of the stem. If covered I think it would go much lower. Mine has yet to flower but I look forward to the 25cm scented white trumpets with great anticipation. (H4)

Trichocereus poco

I bought a good-sized plant of this in 2003 from the late Terry Hewitt and it has spent every year growing outside, including through my three bad winters. It has yet to flower but grows steadily, adding around 2–3cm of stem annually. The 15cm-thick stems are densely covered in long, thin spines, straw-coloured at first, ageing to brown, which seem to shed any moisture falling upon the plant. It has a side branch

Trichocereus chiloensis.

Trichocereus poco.

Trichocereus pachanoi.

and looks absolutely amazing at the base of my south-facing house wall. In habitat it is found in Bolivia and Argentina at extreme altitudes to around 4,200m. (H4/5)

Trichocereus scopulicolus

This is a recent addition, which is similar to the spineless forms of *Trichocereus pachanoi* but fatter.

Trichocereus tarijensis

This is another heavily spined, high-altitude columnar cactus that either is, or is not, the same as *Trichocereus poco*. I have a 1.5m-tall specimen of this that has been growing outside for nearly twenty years, a weather-scarred veteran of my three worst winters. Bought originally labelled as something else, it has flowered three times, a ring of candyfloss pink flowers at the top of the crown confirming the identification. I would offer that, by appearance, it is quite unlike my *Trichocereus poco* in that the stems are far less spiny, much fatter, with fewer, thicker ribs. They may well be the same but they certainly do not look like it. Maybe the *Trichocereus poco* will flower one day by way of confirmation. (H4/5)

Trichocereus tarijensis.

Trichocereus 'Flying Saucer' in a mass of colour.

Watch List

Trichocereus spachianus is very similar to *Trichocereus schickendantzii*, possibly more erect, with fewer ribs and longer spines. I have not grown it for long but so far it looks promising, having made it through my early cold winter. I grew *Trichocereus escayacensis* for several years but it became crowded by other plants, finally weakening and dying. I would like to try again as it is another chunky, tall, columnar cactus with a slightly different vibe about it. (H4)

I have yet to get around to growing *Trichocereus taquimbalensis* or *Trichocereus werdermannianus* (which may or may not be the same as *Trichocereus terscheckii*) but keep my eyes peeled for reasonably priced specimens, as both are hefty and imposing columnar cacti. On a smaller scale, I have recently planted *Trichocereus* 'Flying Saucer', a short-growing columnar cactus of complex hybrid origins with the most stunning huge, bright pink flowers – the most impressive of any cactus I have grown. It came through my early cold winter just fine and is just starting to branch at the base, hopefully forming a chunky clump. (H3/4?)

Basking among the spines.

CHAPTER 6

WOODY LILIES

The term woody lilies is something of a catch-all name that equates approximately with the family Asparagaceae, once referred to as the Liliaceae, and including genera such as *Agave*, *Beschorneria*, *Dasylirion*, *Nolina* and *Yucca*. Plants from this family make up a large proportion of the exotic xerophytes and have a chapter all to themselves.

Agave

Family: Asparagaceae
Subfamily: Agavoideae

Agave is a large genus of rosette-forming succulent plants, generally with spiky leaf margins and tips, distributed throughout the New World from the central USA down to northern South America and across to the West Indies, with a hotspot in Mexico. Exactly how many species and how they sit taxonomically is, of course, a hotly disputed area among experts but most would agree a figure in excess of 200, with hybrids, varieties, forms and cultivars all thrown in for good measure. Individual rosettes are long-lived, reaching anything between 10 and 50 years, then they flower and die, leaving behind offspring in many forms such as seed, offsets and bulbils. They are widely referred to by the indigenous name *maguey*.

The Most Useful of Plants

Humans have had a relationship with *maguey* plants ever since they arrived in the New World, utilising them for food, fibre, fodder and fencing, though it is perhaps as a source of drink that agaves are at their most interesting. As a plant approaches maturity, just before embarking on its final push for flowering, the nascent bud is cut out of the plant and a hollow removed from the centre. Sap that would have otherwise hydraulically propelled the inflorescence skywards, gathers in the internal well from which it is then collected over a period of several months, anything up to 1,500l in total. Drunk fresh it is sweet, with a slightly eggy note, and called *aguamiel*. Left for 2–14 days, depending upon the species, this ferments into a strongly alcoholic beverage called *pulque*, which has a flavour reminiscent of cider but with just a hint of sewage. It has been produced for centuries, if not millennia, as a ready source of cheap intoxication for anyone with access to agave plants. Such was the reverence for the drink that the Aztecs assigned a god to the plant and beverage: Mayahuel. I have tried both but found them not to my taste.

During the post-conquest sixteenth century, the process of distillation was brought to Mexico. It was discovered that cooking a mature agave, trimmed of its leaves then roasted in an open fire-pit, refined the sugars within the plant. The cooked body was then mashed and the resultant juice distilled to give an extremely potent spirit – mescal. Liquor made from the *Agave tequilana* plant, growing in or near the city of Tequila, is said to have greater fragrance and flavour, and is known eponymously as tequila. Once the preserve of local cottage industries, both versions of the drink have become extremely popular worldwide. It is not a quick process – agaves yield the greatest amount of sugars when mature, anything up to fifteen years old, depending upon the species, so with farmed agave plants struggling to keep pace with demand, there are sadly now pressures being brought to bear against wild populations to bolster supplies.

Agave cut for collection of aguamiel, stoppered by a large stone.

OPPOSITE PAGE: *Agave ovatifolia* in habitat.

The sumptuous symmetry of *Agave montana* showing beautiful bud printing.

Some species thrive in the most hostile and arid environments to the extent that they have become noxious weeds outside of their native habitat. Others enjoy moist cloud forests, some in temperate woodland, and it is generally (but not exclusively) these last two provenances that we look towards for species most appropriate for the UK. As garden plants they are captivatingly collectable; maybe it is something about the radial symmetry of the rosettes that appeals, the imposing structural form or maybe flirting with the inherent danger from the spikes. One of the more attractive features of agave plants is the bud printing that is evident on the leaf surfaces. When tightly clasped together in the heart of the plant imprints are left on both leaf surfaces from the teeth or edge of the leaf that was curled so firmly next to it – a bit like waking up in the morning and finding crease marks from the pillow on your face, except that these marks remain for the duration. Whatever the reason, many are among the most desirable, yet enigmatic and frustratingly fickle, plants it is possible to grow in the dry garden.

Keystone Species

Agave bracteosa

Common name: spider agave, squid agave
Distribution: north-east Mexico on vertical limestone cliff faces
Hardiness: H4/5

Agave bracteosa is probably the most atypical member of the genus to kick off this chapter but actually one of the toughest for UK gardens. It grows into an endearingly scruffy 1.5m-diameter rosette of slender, wiggly leaves that have no marginal teeth and a relatively soft tip. The central 'spear' of the plant, where the new leaves emerge, is unusual, too – rather than clasping each other tightly, which is the usual state of affairs, the leaves peel back from the tip almost like pursing lips. Offsets are made generously: my oldest plant has lived on a vertical face of a drystone rock wall for a couple of decades and offsets have found their way out of several crevices among neighbouring plants. It has yet to flower but, when it does, will push out an erect foxtail of creamy white flowers, up to maybe 2m in height.

In habitat the plant chooses to live on sheer limestone rock faces in the Mexican states of Coahuila and Nuevo León, notably in the majestic Huasteca Canyon near Monterrey, often sharing crevices with *Agave victoriae-reginae* and *Hechtia* species, and often on the shadier north side. Given the extreme environment of its home, it is all the more remarkable that *Agave bracteosa* seems so adaptable to many garden situations, with a proven hardiness record in most of the UK. My original plant has sailed through twenty winters, including my three worst, and I have heard accounts of plants surviving –12°C in the open garden without issue.

Agave bracteosa.

Agave bracteosa in habitat on a limestone rockface.

Agave bracteosa has a unique presence and texture that makes a great contrast with other succulent plants. It makes a great container plant, being happy when the roots are constrained. Growing it vertically in a rockery face or drystone wall enhances the shape, allowing the leaves to cascade downwards from the plant as they would in nature, but also ensures they get the perfect drainage they prefer.

Agave filifera

Common name: thread agave
Distribution: central Mexico
Hardiness: H4

Agave filifera is another tough agave that does not really conform to the 'typical' blueprint. Dense rosettes grow to around 60cm across, leaves are narrow, incurving, unarmed at the edges but with curling marginal threads, giving the whole plant a cute, slightly dishevelled look. Sharp terminal spines will, however, remind the careless handler of its agave bona fides. Bud printing on the underside of the leaf is variable and can be particularly striking on some specimens. Offsets are common. Upright foxtail spikes of reddish flowers up to 2.5m high are produced in early summer, although I have yet to witness one in this country on a plant grown outside.

I have seen this in its habitat in the high mountains just north of Pachuca in Hidalgo at Mineral del Chico, at around 3,000m altitude, growing with *Agave salmiana* subsp. *crassispina*, *Furcraea parmentieri* and *Echeveria elegans*, just at the upper edge of the treeline. It gets very cold there during winter, routinely down to –10°C; it was certainly still below freezing at 10.00am the last time I visited there, with cool misty nights throughout the year.

My *Agave schidigera* that masqueraded as *Agave filifera* for nearly 30 years.

Agave filifera in Chris Grundy's garden.

There is massive confusion in the European nursery trade – and, if I am honest, in my mind – between this and the related species *Agave schidigera* and *Agave multifilifera*. I grew a plant, bought as *Agave filifera*, for the best part of 30 years, believing it to be that taxon. Everything seemed correct – the size, width of leaves, marginal filaments, bud printing, plentiful offsets. Then it flowered and the 5m-high inflorescence revealed that it was, in fact, *Agave schidigera*. I have additional plants of *Agave filifera*, *schidigera* and *multifilifera* that, once settled into a life outdoors, appear to be interchangeable. I suppose I will have to wait until they flower to find out for sure. Whatever their identity, they are certainly tough – my *Agave schidigera/filifera/multifilifera* have remained unprotected and undamaged through all of my worst winters.

Agave filifera growing at 3,000m altitude on the peak of El Chico, Hidalgo.

Agave schidigera pushing up its 5m-tall inflorescence.

Beautiful starry flowers of *Agave schidigera*.

Agave gentryi 'Jaws', atypical of the species but familiar in cultivation.

Agave gentryi

Common name: *Maguey Verde*
Distribution: north-eastern Mexico
Hardiness: H4(5?)

Agave gentryi (syn. *Agave macroculmis*) has a wide distribution in nature and, as you would expect, can be of variable appearance. Some forms have narrow, guttered leaves with the 'S'-shaped curve usually associated with *Agave salmiana*. One clone of this form has particularly pronounced teeth and mammillae (swollen bumps along the edge of the leaves, also known as 'teats'). This has been propagated in tissue culture for wholesale distribution under the wonderfully apt cultivar name 'Jaws' and is now probably the most widely grown and familiar face for the species. Yet I would say in nature it is the least common form. More usually plants have short, wide leaves – sometimes guttered, sometimes flat – and often with textured or shallowly dimpled surfaces, as though sculpted from clay and depressions from the artist's fingertips remain. Some forms are very close in appearance to the related species, *Agave montana*, to the point where it is sometimes difficult to tell them apart – in fact they were once considered to be the same taxon (*see* box on p.82). Rosettes can be anything from 80cm across to a whopping 2m. Some have very strong teeth, some are almost toothless, some produce many offsets, some but a few. Yet, somehow, they all identify clearly as *Agave gentryi*. I have yet to hear of one flowering in the UK but in Mexico the inflorescence is a thick, branched, 4m-high club of overlapping bracts through which, eventually, the gorgeous bright yellow flowers emerge. The stalk starts to head upwards in late summer one year, pausing over winter to then resume the following spring, those fleshy bracts going some way to protect the delicate buds from cold while at rest.

Agave gentryi is fairly widespread across north-eastern Mexico, found atop high mountains in the states of Tamaulipas, Nuevo León, Coahuila, Zacatecas and San Luis Potosí, usually between 2,000m and 2,500m but some populations, such as the massive plants at Cerro Potosí in Nuevo León, grow way up to 3,400m. Often they are seen in woodland and often upon open sunny hillsides; always their chunky rosettes

A narrow-leafed form of *Agave gentryi* growing in Darren Hobbs' Newport garden.

A more usual form of *Agave gentryi*, close in appearance to *Agave montana*.

have huge presence. I have had the pleasure of encountering these lovely plants at many different locations in the wild but my favourites are those chunky plants that are scattered on the roadsides and hillsides along Highway 61 from Doctor Arroyo north towards La Ascensión, Nuevo León, near the border with Tamaulipas. Plants resembling these entered the commercial loop erroneously described as *Agave montana* 'Baccarat', later as a 'hybrid' but in reality are as nice an *Agave gentryi* as you could possibly want.

I bought my first *Agave gentryi* in 1996 and it has lived in the garden for nearly all of that time, growing steadily at first but picking up speed in later years. It shrugged off the worst of my winters and has only really been damaged when, in January 2018, storm Eleanor uprooted a polytunnel, which was tossed around the garden a bit before landing smack dab on top of the poor thing, breaking half the leaves. It just shrugged that off too, however, and now you would never know. I have added more and more plants and now have several clones – some bought, some raised from seed. You can still sometimes see plants and seed listed under the old name of *Agave macroculmis* and, for some reason, these seem to be consistently great plants.

Agave gentryi is my absolute favourite of all the *Agave* species I grow, ticking all my boxes – large with interesting form and great presence in the garden, tolerating cold, snow and ice better than nearly all others. I always think back to my travels in Mexico where I have seen it in the wild.

A handsome form of *Agave gentryi* growing near the town of La Escondida, Nuevo León.

Agave 'Baccarat' growing at East Ruston Old Vicarage.

An exceptionally broad-leafed form of *Agave gentryi* growing at 3,400m altitude at Cerro Potosí, Nuevo León.

A Confusing Ménage à Trois?

There are giant agaves living high in the mountains of north-eastern Mexico and a quick read of literature highlights the names *Agave gentryi*, *Agave montana* and *Agave macroculmis* – but what is the relationship between them? We have to delve into the past to unravel this mystery. In 1889 the Italian botanist Agostino Todaro described and illustrated plants growing in the botanic gardens of Palermo in Sicily, one of which he named *Agave macroculmis*.

Fast forward to 1982 and H.S. Gentry, in his massive tome *Agaves of Continental North America*, ascribed Todaro's name *Agave macroculmis* to plants he had seen in north-east Mexico, since they resembled the illustration in Todaro's manuscript. He clearly had doubts, though. Gentry wrote: 'In certain localities, as upon the high mountains east of Saltillo, there is a marked variability in the inflorescence…. Todaro's plate, which represents the type, differs in some respects from the wild plants I am assigning to this name.' Gentry goes on to suggest that a difference in growing conditions at Palermo might explain the different appearance of the inflorescence he noted. He continues: 'If this explanation is not tenable, then the *Agave* assigned here may require a new name, of which we already have far too many in the genus.'

In 1990 this was revisited by German botanist Bernd Ullrich, postulating that the inflorescence in Todaro's plate is, in fact, that of *Agave atrovirens*, again somehow deformed by the Palermo climate. This has consequently led to the name *Agave macroculmis* being dismissed as a synonym for *Agave atrovirens*, leaving the substantial plants that Gentry observed in habitat without a name. Ullrich chose to honour the great man by naming this species *Agave gentryi*, which is how it stands today.

The last piece of the puzzle fell into place when, in 1996, Mexican botanist José Villarreal recognised, as Gentry suspected, that the plants 'upon the high mountains east of Saltillo' are, in fact, distinct from *Agave gentryi* and published a paper splitting these off into the new species *Agave montana*.

Scanned image from Todaro's eighteenth-century manuscript describing *Agave macroculmis*.

Agave lophantha

Common name: none applied
Distribution: eastern Mexico into south-east Texas
Hardiness: H4/5

Agave lophantha (syn. *Agave univittata*) is another one of those widespread, variable plants. There seem to be two main forms: one has slim leaves and grows up to around 90cm diameter; the other has shorter, wider leaves and reaches maybe 60cm across – the actual form of this type being perhaps more familiar as its variegated cultivar 'Quadricolor'. Both have mid-green leaves, usually with a central line of yellowish-green. But it is the teeth that can differ the most: they extrude from the brown, horny leaf margins to varying degrees of toothiness from almost non-existent to maybe 10mm. Some forms sucker like crazy, others are more restrained. Flowers are a foxtail spike of greeny-yellow reaching 4m or so high.

Agave lophantha, narrow-leafed form.

Wide-leafed form of *Agave lophantha* at The World Garden, Lullingstone.

This has a distribution from south-eastern Texas down the eastern side of Mexico to Veracruz. It seems happier in slightly wetter climates, as long as the soil is well drained, favouring rocky places such as cliffs and road cuts. I have seen it in its habitat in several places and it always looks handsome, especially in a moist, subtropical location where the plants have grown huge and lush.

In my garden I have had the narrow-leafed form of this growing happily in a drystone-wall rock face for nearly twenty years and it has never been damaged by the weather, taking my worst winters in its stride. I know of people who have grown this through –10°C in the open garden and I am sure it would take much lower temperatures if covered.

There is an extreme form with exceptionally pronounced, scary-looking, rip-saw teeth that is sometimes referred to as *Agave heteracantha* in Europe (but not in the USA). The jury is out as to whether this is a separate species from *Agave lophantha*; certainly in flower they are virtually indistinguishable, but it is definitely not *Agave xylonacantha*, a name that is often wrongly applied but a different beast entirely. I have grown this form for many years through my worst winter

Agave lophantha growing on a limestone cliff in Tamaulipas, Mexico.

Very toothy form of *Agave lophantha*, sometimes known as *Agave heteracantha*.

and late winter but it did not take kindly to the early winter and was actually quite badly damaged – in contrast to *Agave lophantha*. An ostensibly similar species is *Agave lechuguilla*, which resembles the narrow-leafed form but is smaller, more rigid, with tougher leaves displaying dark dashes on the underside. In the wild this is a weedy plant, offsetting like crazy and covering hectares of parched ground, making progress on foot rather tricky. To my eyes it is not massively ornamental but is, however, extremely tolerant of cold and extreme drought.

Agave montana

Common name: mountain agave
Distribution: north-eastern and central Mexico
Hardiness: H4(5?)

Agave montana is one of the most beautiful agaves to my eyes. The chunky, almost always solitary, rosettes are up to 1.5m across, densely packed full of short, wide but thin, incurving leaves that, overall, give the impression of a houseleek on steroids. Individual plants can vary considerably even within the same population but, generally, leaves are deep, glossy green, sometimes with a glaucous patina and almost always beautifully marked with clear, chalky, bud printing. Marginal teeth are usually reddish-brown. Some are greyer than others, some have less vivid bud printing, some have more pronounced teeth, but they are all marvellous. At maturity the rosettes will carry over 100 leaves. The first UK seed-raised plants have now flowered at seventeen years old, producing a thick, club-like spike – covered in fleshy, purple bracts – that starts to emerge in late summer the first year, pausing over winter, then continuing in spring to reach, ultimately, around 4m high. Short side branches extend and buds start to open during the early and mid-summer with bright yellow flowers, appearing initially as tufts sticking out of the bracts. If offsets are ever made, it is generally at this point, once the flowering is under way but, usually, flowering is the final act. With the onset of flowering, and with no further need to photosynthesize, the leaves do not seem to bother replenishing their chlorophyll, allowing hues of yellow and red to appear, adding to this wonderfully colourful swansong.

Agave montana in Rob Rudge's Taunton garden.

Agave montana comes from very high altitudes, between 2,800m and 3,200m, at the top of just a few mountains in the north-east of Mexico. There is an accessible population at Cerro Peña Nevada, a mountain straddling the states of Tamaulipas and Nuevo León near the town of Miquihuana. I have visited this site on four occasions, spending many, many hours wandering around the plants, just soaking up their beauty and variability, whilst waiting for my friend Mark to return from seeking out dahlias. Some of these plants are almost indistinguishable from some forms of *Agave gentryi* I have seen, underscoring their close relationship – although, perhaps unexpectedly, not the population that shares the lower slopes of this same mountain, which display more narrow, guttered leaves. Bizarrely, there is a disjunct population of *Agave montana* nearly 500km further south on the high peak of Cerro La Laja in Querétaro, in central Mexico. These plants are similar, although stubbier with shorter, wider leaves and, on occasion, do actually make one or two offsets. These high peaks are cold and snowy in winter, cool and misty in summer. In fact my first attempt to visit these southern plants had to be aborted

Twin towers of *Agave montana* flowers at East Ruston Old Vicarage.

Agave montana in Darren Hobbs' Newport garden showing colourful changes when flowering.

Forest habitat of *Agave montana* at 3,000m altitude on Cerro Peña Nevada, Tamaulipas.

The perfect *Agave montana* at Cerro Peña Nevada.

Chunkier *Agave montana* plant from population growing on Cerro La Laja, Querétaro.

An Internet Romance

Paying homage to *Agave montana* in the flesh was the reason I first made a trip to Mexico back in 2004. A year or so earlier I had got chatting on the internet with a Texan guy, Richard Travis, who has travelled extensively throughout north-east Mexico and he showed me pictures of the now familiar squat rosettes of this beauty. It really was love at first sight. Rich said, 'It really is cold, damp and horrible up the north-east, you Brits really ought to go up and see what is going on there.' It took a while to arrange the details but, together with my friend Toby Shobbrook, I made it there in November 2004 and met up with Richard who kindly spent a fortnight showing us where all the fascinating plants were. I lost a little bit of my heart to Mexico on that trip, the first of eight, during which I covered a lot of territory, saw a lot of plants, ate a lot of tacos and drank a lot of *michelada*.

because it was so misty we could not see the road.

With such a high mountain provenance with cool, temperate growing conditions, it should be safe to assume that they would be perfect to grow in cool, temperate UK gardens. Right up until my bad early winter I would have bet the house on them being undamaged by anything the English winters could throw at them. Unfortunately, the strange combination of a long, wet and mild autumn, in which the plants were still actively growing, followed by a sudden onset of freezing, left these plants vulnerable. When dormant they will take down to at least –10°C with no issues.

Agave ovatifolia

Common name: whale tongue agave
Distribution: Sierra Lampazos, north-eastern Mexico
Hardiness: H5

A superlative example of *Agave ovatifolia* in Rob Stacewicz's garden.

Agave ovatifolia is a stunningly attractive plant with ghostly pale, glaucous, silver-blue rosettes around 1.5m across, rarely 2m, jam-packed with short, wide leaves that are varyingly armed with black marginal teeth and leaf-tip spines. It is the form of the leaf that makes this species so appealing for me: the centre dips into a shallow bowl but the edges and tip curl away from this, fancifully resembling a whale's tongue, hence the common name, but to me more reminiscent of a hybrid tea-rose flower as it begins to open. Added to this are extra three-dimensional effects when the leaf surfaces on some older plants develop the most attractive pleating and dimpling. In nature it is almost always solitary but in cultivation it seems a few offsets are produced on some plants. I have recently added a group of five seed-raised youngsters to an area of new planting and four of them have started offsetting already. Greenish-yellow flowers, when they appear, are borne on a thick 'Christmas tree' panicle that reaches 4m or so. In the USA this takes just ten or so years from seed, very rapidly for such a large agave, so it is a real 'live fast, die young' plant.

Closer look at the crown of *Agave ovatifolia* in Darren Hobbs' Newport garden.

Breathtaking specimen of *Agave ovatifolia* in Rob Rudge's Taunton garden.

Agave ovatifolia 'Vanzie' growing in Ed Croft's East Sussex garden.

In nature these are found growing in just one place: Sierra Lampazos in Nuevo León, Mexico, north of Monterrey, on what is effectively a limestone pavement, up to an altitude of around 1,200m. Sadly I have not been there and understand it is becoming increasingly difficult to access these wonderful plants as it is necessary to travel through closed, privately owned land. Similar plants have been found elsewhere further south, *Agave* 'Huasteca Giant' being one, but their identities have yet to be established. If they do end up being the same taxon, which I personally doubt, they certainly do not have the same majesty as these Sierra Lampazos plants.

A relative newcomer, *Agave ovatifolia* was first introduced into limited cultivation in the USA in the 1980s but was not formally described until 2002, material reaching the UK and Europe in the mid-2000s. Since that time it has proved to be a superlative garden plant. It takes frost, cold, ice and snow better than most of the larger agaves, with reports of it having withstood temperatures down to −12°C unprotected. It also grows more rapidly and is less xeric than most agaves, not seemingly too fussy about winter wet and actually performing better in wetter parts of the UK than in my dry garden. It is rare for a plant that arrives into the marketplace with such a fanfare to live up to the hype but, in this case, it is entirely justified. It is hard to overstate the importance of this species as a key plant for UK xeric gardens and, speaking personally, I suspect it is in danger of usurping my beloved *Agave gentryi* for the top spot in my affections.

One of the *Agave ovatifolia* imposters in Ed Croft's garden.

Agave parryi

Common name: Parry's agave
Distribution: Arizona, New Mexico, Texas, north-western Mexico from Sonora to Zacatecas
Hardiness: H5/6

Agave parryi (syn. *Agave chihuahuana*, *Agave patonii*) has an enormous range in nature and, largely based upon geographical population locations, has been divided by taxonomists into two subspecies: *Agave parryi* subsp. *neomexicana* (syn. *Agave neomexicana*) and *Agave parryi* subsp. *parryi*, from which three varieties have been recognised: var. *cousei*, var. *huachucensis* and var. *truncata*. There is much crossover in appearance between different populations, which has led, in the UK and Europe at least, to a fair amount of confusion as to what is representative of each variety.

As far as 'typical' appearance goes, a default starting point would be *Agave parryi* subsp. *parryi*, which has rosettes that are dense globes, maybe 80cm across, of silver-grey, wide, incurving and relatively flat leaves that have black marginal teeth and terminal spines: the overall effect often described as resembling a globe artichoke. Offsets are usually numerous. Stout flowering stems, when the time comes, can reach up to 6m and bright golden flowers open from colourful reddish buds. This comes from Arizona, New Mexico and north-western Mexico at anything up to 2,800m alti-

Agave parryi subsp. *parryi* in Ali Kent's garden.

tude and in habitat it is hardy to around −10°C or lower. Named cultivars have been selected, including 'J.C. Roulston', with redder teeth and said to be better for moister climates.

The var. *couesii* has rosettes that are smaller and more open with leaves that are generally narrower and not armed with as many teeth. This is from the mountains of central Arizona and one of the hardiest types, taking down to −18°C in habitat and cultivation in the USA.

In var. *huachucensis* the rosettes grow to be a bit larger, to 90cm, with wider leaves. This hails from south-eastern Arizona (named after the Huachuca Mountains there) and across the border into Sonora and Chihuahua and is not as hardy, maybe to −10°C. It is an exceedingly impressive plant with large and densely packed rosettes, doing the 'globe artichoke' thing quite nicely.

In var. *truncata* the rosettes are a little smaller, to maybe 75cm across and, typically, the leaves are much shorter with the terminal spine appearing from a truncated, blunt leaf end. This is a Mexican plant from the Durango/Zacatecas borders and said to be one of the hardier forms, taking down to −12°C. A clone grown in the Huntingdon Botanical

Agave parryi var. *couesii* in the Desert Botanic Garden, Arizona.

Pot-grown *Agave parryi* var. *huachucensis* at an Arizona nursery.

Agave parryi subsp. *parryi* 'J.C. Roulston' in Chris Grundy's garden.

Agave parryi var. *truncata* at an Arizona nursery.

Agave parryi subsp. *neomexicana* in a Gran Canaria garden.

Gardens in California is thought to be the most desirable of this variety.

Plants of subsp. *neomexicana* are similar to var. *couesii* above in that they are smaller, with more open rosettes comprising narrower leaves. I would say these are more striking as, generally, they are a paler silvery-grey in colour, slightly guttered, with very pronounced back marginal teeth and terminal spine, giving a strong contrast with the pale leaves. This form comes from New Mexico and is probably the hardiest of all the *parryi* group, taking temperatures down to maybe –20°C in its habitat.

All of that is in theory. In practice things are a mess. I have previously grown various plants that have been assigned all sorts of names within the *parryi* group that do not conform to the 'official' line and, frankly, most of these have generally struggled to do well. I have one clone, arriving with just the name *Agave parryi*, that has been exceptional, and I know of a few other people who have grown the same plant from the same source and been equally satisfied with its performance. I would imagine it to be *Agave parryi* subsp. *parryi* as it does not quite fit comfortably with the others. This has been through all my worst winters unprotected, despite a bad start – eighteen years ago when I knocked it out of the pot to plant it, most of the roots stayed in the pot in the claggy compost in which it had been growing. I used to grow a plant labelled as *Agave parryi* 'Buenaventura', which I guess refers to where the seed was collected as there is a town named Buenaventura in Chihuahua. The plant was gorgeous, having a very slight pink tone to the grey and very black but reduced teeth. It went through a few sharp winters as a tiny plant with no issues so I decided to move it to a premium spot in the garden and, of course, killed it in the process. I currently have a plant that would conform to var. *truncata* that is establishing quite well and has taken –6°C without damage, with more types biding their time in my greenhouse waiting to be planted out.

I am keen to try more of the *parryi* group as I think the striking combination of ghostly silver-grey rosettes with black teeth and spines is stunningly attractive. They are more compact than some of the other similarly coloured *Agave* species and, if a good clone can be unearthed, are virtually impervious to the UK winters. Certainly my good subsp. *parryi* clone is one of the nicest agaves I have ever grown. The trick is knowing what you have got but, without the continuity of seed provenance to plant sale, that is not easy.

It is probably worth mentioning *Agave* × *gracilipes*, a naturally occurring hybrid between *Agave parryi* subsp. *neomexicana* and *Agave lechugilla*, found where their ranges overlap. Plants are beautiful, compact, silver-blue rosettes of narrow leaves armed with black teeth and spines, almost like a hardy version of the tender *Agave macroacantha*. The leaf reverse inherits dark striations from *Agave lechuguilla* and is most attractive. I grew one for several years here until it flowered.

Agave × *gracilipes*.

Agave striata

Common name: needle agave, *Espadin*, *Sotolito*
Distribution: eastern Mexico as far south as Mexico City
Hardiness: H5/6

Needle agave is a great common name for *Agave striata* as it really does look like a great spiky 60cm globe of curved knitting needles and is about as dangerous. Slender, angled, stiff, rough-textured and sharply tipped leaves are packed into a dense dome, which, in nature, offsets profusely to form extended colonies that can reach 3m or more across. Plants are usually green, certainly when growing here in the UK, but when drought-stressed there are blue and red forms to be seen, even within the same population. The flower spike is a 1.5m-long foxtail bearing, in late summer, red flowers opening in succession from around halfway up. After flowering, *Agave striata* departs from the usual agave model by making new offsets within the leaves and, in time, the plant becomes a multi-headed clump.

Agave striata in habitat, northern Mexico.

Agave striata sporting two fine flower spikes.

Agave striata is a very common plant along the eastern side of Mexico from just below the USA border to around the centre just above Mexico City. It generally grows into massive colonies in very dry, flat, scrubby locations with *Opuntia* species and other extremely xeric plants, not usually at cooler high altitudes, which makes it all the more surprising that it actually grows well in the UK.

I have had a plant of this growing in my rock garden for nearly twenty years during which time it has never been damaged. In that time it has flowered five times – once with three spikes in the same year – and it is now a fairly dense clump of dangerously impenetrable spikes. Fallen leaf litter has to stay put as I have all too often prodded myself on the leaf tips. It is almost as though they have a hand magnet on them as, whenever I get within reach, I seem to stab myself no matter how careful my approach. There is a subsp. *falcata* that is smaller with more tightly curved leaves that I have grown but lost. Based on that sample of just one I would say it is less hardy, but do not quote me on that.

Agave stricta is an extremely similar-looking plant from a long way further south in Oaxaca and Puebla in Mexico that grows in almost identical situations and in almost identical ways. Given its southerly provenance it should not really be hardy, yet I have a plant of this name that has survived outside in my garden for nearly twenty years. With that said, away from Mexico I find them hard to tell apart, so perhaps this inability was shared by the seller all those years ago. It is a much smaller plant that has never flowered.

Agave stricta in habitat, southern Mexico.

Worth Considering

Agave havardiana

A chunky species from western Texas and the north-eastern states of Chihuahua and Coahuila in Mexico, found growing on rocky grassland slopes, this has broad, stiff leaves of pale silver-grey, trimmed with black teeth and spines, carried in

Agave havardiana in the Desert Botanic Garden, Arizona.

dense rosettes to 1.5m across. It is a handsome agave that is extremely hardy, taking down to –25°C in some parts of its range. Some populations are stunningly attractive, with really wide leaves and strong teeth and spines, such as those found in the Davis Mountains, and are as beautiful as any large agave can be. As far as I know it has not been widely grown in the UK. A friend grew one in his sunny front garden not far from here for many years, which grew into a decent-sized plant until it had to be removed. Yet until very recently I have never grown it. (H6)

Agave mitis

(syn. *Agave celsii*)

These are really cute agaves that grow to around 90cm across with fairly open rosettes of slightly brittle leaves that can be anything from shiny Granny Smith apple green to chalky white (the glaucous form was once given varietal status var. *albidior*, but it is now considered as just part of a varied species). Leaf margins are inconspicuously toothed and there is a softish terminal spine. They grow almost exclusively on cliffs and rock cuts in subtropical cloud forest-type environments where rainfall is much higher, often in shade, at altitudes up to around 2,200m, although I have seen a population growing way up high at 3,100m in the open at Cerro La Laja in Querétaro. As they age, offsets are produced inside the rosette in leaf axils and they can cluster into quite big clumps given time and space. They are not the hardiest of plants, I have had them damaged at –6°C and lost them at –8°C but for the wetter, milder south-west of these isles they are a fabulous and reliable plant.

Agave obscura

(syn. *Agave polyacantha* var. *xalapensis*) is very similar to look at and it is difficult to tell the two species apart. In nature it is easy as they do not grow together: this has a fairly limited distribution in eastern Mexico in Veracruz growing upright on lava fields and woodland floor. I have found it to be a teeny bit hardier than *Agave mitis*. (H4)

Agave mitis on my small cliff, just before my worst winter.

Agave parrasana

This is a compact and beautifully marked hardy species, said to withstand down to –15°C cultivated in the USA. Some forms have very strong spination and bud printing, and at maturity the rosettes are densely packed with firm, short, wide leaves: the common name of cabbage head agave giving a clue as to the look. Most often plants remain solitary; some are prolific offsetters. My friend, Jan Kolendo, grew this in his Surrey garden for ten years or so, where it started as a small plant and

Agave obscura.

An impressively toothy *Agave parrasana*.

Agave parrasana in Jan Kolendo's Surrey garden.

eventually made it through to flowering. As with a few other montane agaves, flowering in this one starts late one summer, pauses over winter, then continues into the next summer when the thick 4m spike, covered in fleshy purple bracts, is adorned with tight bunches of yellow flowers. In the wild, distribution is limited to a few mountains in Coahuila, Mexico. (H4/5)

Agave polianthiflora

One of the smallest agaves, this is also extremely hardy. It makes spiky little

The unusual flowers of *Agave polianthiflora*.

rosettes, to around 25cm across, comprising narrow, toothless but sharp-tipped leaves that are 'D'-shaped in cross-section, giving them great rigidity. Both leaf surfaces are beautifully marked with chalky bud printing and have conspicuous white marginal filaments. They really are cute little things but precocious, too, as they will often flower by their tenth birthday. And it is the flowers that set this agave apart, being full red tubes and unlike any other I have seen in flower. It comes from the Sonora/Chihuahua borders in northern Mexico and is quite xeric in its requirements but, if dry, can take down to –15°C. (H5)

Agave salmiana

This is one of the commonest large agaves in central Mexico, naturally growing as far north as San Luis Potosí and then south into Puebla throughout most of the high, dry central plateau. It has been cultivated since prehistoric times as a source of fodder, boundaries, soil-stabilising barriers and fibre, most of these being ancillary to its principal use for *pulque* production. As a result it has been spread far and wide and it is sometimes hard to know where cultivated plants end and wild plants

Agave salmiana subsp. *crassispina* flowers close up and colourful.

A chunky *Agave salmiana* subsp. *crassispina*.

begin. In the more remote parts of its range it tends to take on a particular form – quite compact (by *Agave salmiana* standards) at 1.6m across, not particularly toothy, thick chunky and broad leaves, and can be seen at high-altitude spots like Mineral del Chico in Hidalgo, El Doctor or Cerro La Laja in Querétaro, way up to 3,100m altitude. These splendidly chunky plants are *Agave salmiana* subsp. *crassispina* and, smaller size notwithstanding, make the best garden plants for the UK, being a touch hardier than most other forms. I have grown several clones of this for many years; recently my oldest plant flowered spectacularly at around 30 years old. If dry they can take below –10°C but resent the freeze/thaw of wet snow for any length of time, so best grown with a cover. This form is sometimes seen offered under its old name *Agave coarctata*. There is a wonderful named variety *Agave salmiana* var. *ferox* that was described from plants growing in La Mortola gardens in Italy that has never since been found in the wild. It is quite distinct with satin, dimpled, mid-green, broad, elegantly recurved leaves possessed of ferocious marginal teeth but, sadly, it is not particularly hardy, usually being badly damaged around –5°C, if that. Unfortunately there are a lot of imposters as pretty much any wide-leafed and toothy form of *Agave salmiana* gets labelled as var. *ferox*. (H4/5)

Agave salmiana var. *ferox* unexpectedly thriving in Ed Croft's East Sussex garden.

Agave salmiana subsp. *crassispina* flowering in my garden.

Agave victoriae-reginae

A special plant, quite different from most other *Agave* species, that is quite variable in the details but always immediately recognisable. This is a compact plant growing typically up to 60cm across and ultimately forms an incredibly dense football-like globe of short, stout, sturdy, three-sided leaves arranged in the most beautiful spiral. Individual leaves are among the most brightly marked in the genus with an almost arts and crafts' design bud-printed in chalky white that appear to have been hand-stencilled on to both leaf surfaces. There are no marginal teeth but a hard, sometimes white, margin. Leaf tips end in an exceedingly hard and sharp spine. Flowers, when they happen, are anything from yellow to purple, held on an outrageously tall 5m foxtail stalk. In nature it grows on limestone – vertically in cliff faces or more horizontally lower down – in scattered locations around the north-east of Mexico, notably in the Huasteca Canyon along with *Agave bracteosa* and even *Agave albopilosa*, if you know where to look (and I did not). I have read of it being hardy down to –15°C in cultivation in the USA but of the many plants I have tried, only one has done well, taking very little damage in my early cold winter. Surprisingly it is one growing upright rather than in my rock face. What I take from this is that with no provenance details available, it is just pot luck. (H4/5)

Agave victoriae-reginae in my garden.

Agave victoriae-reginae in the Huasteca Canyon, Nuevo León.

Agave utahensis

This is a variable complex of plants found growing on limestone in the states of Utah, Arizona, Nevada and California. Taxonomically the species is split geographically into two subsp. *kaibabensis* and subsp. *utahensis*, the latter having two varieties var. *nevadensis* and var. *eborispina*, all of which vary to a greater or lesser degree in size, width of leaf, and elongation of teeth and terminal spines. Of particular note are the larger rosettes of subsp. *kaibabensis*, the extra-long terminal spine of var. *eborispina* and pale blue chunkier shorter leaves of var. *nevadensis*. In their habitat most of these populations grow in extremely xeric conditions, no more than 300mm of rainfall a year, usually less, and also extremely cold in winter, down to –20°C. Much colder than it has ever reached here, yet I have never, once, managed to overwinter any of these outside, despite numerous attempts. They are very slow to grow in the UK and to get one to a size that I consider safe to plant out takes many years, so it is a long and expensive business! I have heard of one or two people who have managed to keep some forms of this going, particularly var. *nevadensis*. I have a theory or two as to why, touched upon in earlier chapters. As a pot plant that is overwintered under glass before it gets the opportunity to get wet in autumn, I think these compact plants score heavily. Perhaps they would succeed better outside if they were reliably and regularly covered in autumn; yet I have spoken with someone in Scotland who grows one in a rockery outside unprotected. One of the many enigmas of this fascinating genus.

Another sacrificial *Agave utahensis* var. *nevadensis*.

Watch List

Agave albopilosa

Very much one of the new kids on the block, this agave was only formally described in 2007. It comes from just one place in the wild, Huasteca Canyon in Nuevo León, the precise location being kept under wraps as it has become extremely desirable for collectors. This is no wonder, it is just so unfeasibly sweet with dense little rosettes of tough little leaves that are unarmed at the sides, have a sharp spine at the tip together with the cutest frayed tuft. So it was with great trepidation that I decided to gamble with my one little specimen outside, giving it pride of place in my south-facing rock wall. I reasoned that it grows alongside *Agave bracteosa* and *Agave victoriae-reginae* so, therefore, it must surely be hardy. After all, it looks tough… and so it has proved to be, taking my early onset cold and late onset cold without any damage at all.

Agave albopilosa in my small cliff.

Darren Hobbs' magnificent *Agave albopilosa*.

Agave atrovirens var. *mirabilis* in the small village of Las Vigas, Veracruz.

Agave atrovirens

On paper this giant agave should be one of the hardiest, living between 2,000m and 3,000m altitude on nearly all the high peaks in the south of Mexico. Coming from a cloud forest environment it is also more moisture-tolerant than most. In practice it seems to resent snow and ice, and gets damaged if uncovered at around –5°C, but for the milder, wetter parts of these isles it is unrivalled for its stature and presence. A particularly desirable form, var. *mirabilis*, is not known in the wild and can only be seen in the small village of Las Vigas in Veracruz. It is just huge and glaucous blue and gorgeous. Tracking one down is another matter, all but one of these I have ever seen in cultivation are imposters.

Agave franzosinii

This is another giant agave that grows into a huge, silvery-white specimen, very much like an improved *Agave americana* in looks but with leaves that are wider in the top third, which have a tendency to curl inwards under their own weight. It does not sucker as freely, either, which makes it a much more desirable plant. Grown with a cover to keep the weather off this is doing rather nicely in Melissa Scott's garden just outside Norwich in Norfolk, taking –10°C dry.

Agave geminiflora

Another atypical agave, this one makes a flexible dome of softish narrow leaves with, on the best forms, curling marginal leaf fibres resembling a larger, more user-friendly version of *Agave striata*. I have not seen this in its habitat but my friend

Agave geminiflora on St Michael's Mount, west Cornwall.

and nurseryman, Steve Mules, grew one to perfection in his south-west Cornwall garden, where it enjoyed the mild, wet climate and eventually flowered. My own plant did not make it through my early cold winter, falling apart after a low temperature of –6°C, but having struggled miserably in my dry climate up until then.

Agave weberi

Another slightly more user-friendly agave with no (or only a few) marginal teeth: this has a lovely tidy form, although it does get quite large. Paddle-shaped, narrow-tipped leaves in a lovely pale blue-grey form a fairly open rosette to maybe 1.8m across. Unprotected from winter weather this struggles to take below –5°C but kept dry it happily goes much lower, being grown from a smallish plant to flowering size by my friend Alison Kent in her north Essex garden, taking down to –15°C under a polytunnel in the worst winter.

A robust *Agave weberi* in Ali Kent's former north Essex garden.

Agave wocomahi.

Agave wocomahi

A different agave, not seen that frequently, this one will make nice toothy rosettes of glaucous grey-green, dimpled leaves with heavy marginal teeth up to around 1.5m across. It did not enter my radar until around ten years ago when a grower in western Germany reported its cold-hardiness. I have grown it through my bad late winter with a cover with absolutely no damage. I left it uncovered during my early bad winter, which resulted in a fair bit of damage to the older leaves, although subsequent growth has been fine. So I would say this one definitely benefits from a winter cover, when I would think it would take down to –10°C without any problems.

Agave × nigra

(syn. *Agave* 'Sharkskin') A naturally occurring hybrid between *Agave nickelsiae* (syn. *Agave ferdinandi-regis*) and *Agave asperrima* (syn. *Agave scabra*) found on just one rather arid hillside in north-eastern Mexico near the town of Saltillo. It could be argued that this has inherited many of the best features of both parents and it is extremely popular with its tough, rough-textured, grey leaves, three-angled and rigid, with jet-black terminal spines and, sometimes, black leaf margins at the top of the leaves. Open rosettes grow to around 1m across. In the USA this is regarded as hardy down to –12°C, yet in the UK it has never performed very well. I have lost three perfectly good plants at nothing like those low temperatures with a new attempt showing no damage after my early winter in a protected spot. I suspect it is another one that would do well either in the milder parts of the UK or under a rain cover.

The Most Virulent of Weeds

Perhaps an obvious omission, I have left *Agave americana* out of the recommended species. Having travelled quite a bit in the Mediterranean, Canary Islands, etc., I find it hard to see it as anything but a terrible weed that, along with some *Opuntia* species, is rampaging out of control through hectares of wild land. So I have placed it here in its own box in the hope that it does not escape into the rest of the book. For me it has little merit in a garden situation: a solitary rosette grows huge, over 2m across – but it does not stop there. Plants constantly produce offsets, scores of them, each of which aspire to becoming 2m monsters themselves. It is not uncommon to find one of these succulent cuckoos wedged into a tiny front garden, getting tattier with each winter yet persistently refusing to be killed off by anything other than extreme cold. If nature is allowed to take its course, at least there is the final glory of the massively impressive inflorescence to enjoy. All too soon, however, the time comes to remove it: a mammoth task exacerbated by irritant sap, chainsaw blunting fibres within the leaves and roots like steel cables. Nevertheless the species, and its gaudily variegated cultivars, remain the most popular agaves on the market.

Agave americana destroying the Gran Canarian landscape.

Agave × nigra in habitat east of Saltillo.

Agave × nigra.

Beschorneria

Family: Asparagaceae
Subfamily: Agavoideae

Members of the genus *Beschorneria* are succulent plants that are closely related to *Agave* and *Yucca*. With one exception they are all trunkless plants forming a rosette of succulent but soft, sword-shaped leaves from the centre of which, after just a few years, is produced a huge, unfeasibly flamboyant, branched flower spike from which dangle scores of tubular bi-coloured flowers. The rosette that has flowered dies but the underground part of the plant throws up offsets resulting, over time, in a clump composed of multiple rosettes. Propagation is by seed, which is readily set in the UK, or division of large clumps.

Keystone Species

Beschorneria septentrionalis

Common name: red false agave
Distribution: north-eastern Mexico
Hardiness: H4/5

Beschorneria septentrionalis has open rosettes of 70cm long, smooth, leathery, shiny apple-green leaves, finely serrated at the margins, broad for their length at 9–11cm and often rippled. Startling radish-red stems emerge straight and continue to grow more or less erect to around 2m or so, branching only towards the lower third, with branches reaching 30cm. Eventually the weight of flowers causes it to tilt slightly from upright. Bracts are the same radish-red but much smaller and less persistent than in *Beschorneria yuccoides*, withering as the stalk starts to unfold. Flowers are tubular but with a closed-in tip, smooth, bright red with the tips seemingly dipped in kiwi-fruit green.

It comes from the north-east of Mexico. I have seen it growing on well-drained rock banks in moist woodland at El Parque Ecológico Los Troncones, Tamaulipas, in surprisingly dense shade along with ferns and the clumping palm *Chamaedorea microspadix*. Given this provenance, in theory it should prove to be both hardy to cold and tolerant of wet, yet sometimes success seems elusive. Where happy it grows extremely well and in some gardens has survived down to –12°C, perhaps tolerating even lower temperatures if covered for winter.

In my garden it has survived all of my worst winters without damage and I have several clones of this in various parts of the garden. The bright green rosettes make a pleasing contrast to the frequent glaucous greys and, in flower, the intense radish-red is a welcome blast of colour. I find it will most usefully take a little light shade too.

Beschorneria septentrionalis.

Beschorneria septentrionalis flowers.

Beschorneria yuccoides

Common name: Mexican lily, *Amole*
Distribution: central Mexico
Hardiness: H4/5

On paper, at least, this species has been split into two subspecies, *Beschorneria yuccoides* subsp. *yuccoides* (syn. *Beschorneria bracteata*) and *Beschorneria yuccoides* subsp. *dekosteriana* (syn. *Beschorneria dekosteriana*), based on geographical and floral differences. In practice the situation is not so cut and dried, so I simply refer to all of them as *Beschorneria yuccoides* until convinced otherwise.

Beschorneria yuccoides typically has a sparse, open rosette of grey-green, sword-shaped leaves, 85–90cm long, widest around the upper third and tapering to a soft tip. The lower surface (and sometimes the upper surface) is rough to the touch with both surfaces generally covered to a varying degree with a glaucous patina. The edges are finely serrated. After a few years the plant will be ready to flower: a spectacular event that starts as a bud nestled within the leaves early in spring. This bud extends into a flower stalk, the tip bending over as it emerges to give a hockey-stick effect, eventually deepening in colour to a strong rhubarb-red with large, fleshy bracts of the same hue clasping the sides. Soon the inflorescence reaches its maximum size – to 3m or so and widely branched – from which dangle scores of rhubarb-red and pear-green flowers, the weight of which often causes the spike to gently tilt to the horizontal. Occasionally old rosettes of plants can show a small above-ground caudex, especially where plants have been trimmed of tatty older leaves, but this is not that common.

In its habitat around central and eastern Mexico, *Beschorneria yuccoides* is found at high altitudes between 2,600m and 3,400m in the cloud forest zone of temperate woodland, generally growing in pockets of organic soil on steep, rocky outcrops in the light shade of the tall tree canopy but also in open, sunnier positions. Rainfall in this environment is surprisingly high – these are not particularly

The 3m-long inflorescence of *Beschorneria yuccoides*.

A form of *Beschorneria yuccoides* at RHS Hyde Hall with a shorter, unbranched inflorescence.

Beschorneria yuccoides flowers.

Beschorneria yuccoides in habitat, central Mexico.

xeric in their requirements – but drainage seems to be very important.

We can take hints from this to aid its cultivation in the UK. A position in full sun or even light shade is fine, with good soil that is well drained. In the open garden, plants are generally hardy to around –10°C. If grown under an evergreen canopy, or given overhead winter protection, they can survive temperatures down to –15°C or so. My plants have taken all my worst winters with no damage.

Young plants are easily confused with young plants of *Furcraea parmentieri*. Typically the leaves of *Furcraea* are more linear with roughness on both surfaces but this is not consistent. According to my friend and nurseryman, Steve Mules, the only sure way is to watch snails and

Beschorneria septentrionalis × *Beschorneria yuccoides* growing at RHS Wisley.

slugs, which will quickly rasp unsightly patches from the leaves of *Beschorneria yuccoides* yet leave the furcraeas alone. Of course within a few years the difference is apparent as the *Beschorneria* will flower colourfully and not, in the usual course of events, form a trunk. *Furcraea parmentieri* will continue growing a trunk for some years before flowering. Ultimately *Furcraea parmentieri* is a good deal more tender and a hard frost much below –6°C is usually enough to see it off in the open garden.

I grow a hybrid *Beschorneria septentrionalis* × *yuccoides* made by a friend that seems to combine many of the best features of each parent, being both compact and colourful. The leaves are greener than those of *Beschorneria yuccoides* and bear some light silvery patin: 50/50 blood-red and apple-green flowers are borne on blood-red 2m stalks, sparsely branched near the middle. A similar hybrid is growing against a wall at RHS Wisley in Surrey.

Worth Considering

The next three species seem to make a natural group, all sharing common features such as narrower, shorter, keeled leaves and a more erect flower stalk. I am less than confident about the true identities of these particular plants, however, and suspect that the original labels might be wrong. Nevertheless I will forge ahead regardless using the names I have come to associate with these plants, airing my doubts as a footnote.

Beschorneria calcicola

In many ways it resembles an even more compact version of *Beschorneria rigida* with mid-green glossy leaves, 40 × 2.2cm, deeply keeled with a V-shaped cross-section and fine marginal serrations. In flower it has a similarly unbranched, erect flower spike of a strong carmine-red to around 1.5m high. Bracts and peduncles are also carmine. The flowers – smaller than those of the other species – are tubular, very slightly hairy, tipped with pear-green and attractively mottled. Despite an unpromising southern Mexico provenance, mine has taken my early and late cold winters without damage. (H3/4)

Beschorneria calcicola.

Beschorneria calcicola flower.

Beschorneria rigida

(syn. *Beschorneria roseana*)

This one is smaller in stature than *Beschorneria tubiflora*, vigorous and outstandingly showy in flower. The leaves are deep glossy green, 60cm long by 2.5cm wide, V-shaped in the centre, narrowly tapering to a slender point and with fine

Beschorneria rigida.

Beschorneria rigida flower.

marginal serrations. In just its second season after planting, it threw up a spectacular erect, unbranched, 2m-high flower spike in a brilliant carmine-red. Strongly coloured carmine and apple-green tubular bells are large for the size of the plant and, like those of *Beschorneria tubiflora*, noticeably hairy. I have grown this for nearly ten years and it has taken my early and late cold winters with no issues. I suspect it would take down to around –10°C just fine, lower if covered. With just a few more successful years growing under its belt this could easily become my favourite in the genus due to its compact nature and bright flowers. (H4/5)

Beschorneria tubiflora
(syn. *Beschorneria tonelii*)

This species has narrow, grey-green, almost linear leaves with a strong V cross-section and bearing a silvery patina, growing to around 50cm long by 3cm wide. They are coarse to the touch on both surfaces with slight marginal serrations. The more or less erect flower spike emerges a fairly dark brownish-red, eventually losing the colour as it extends (to 1.2m). Side branches are few but retain the colour, as do the bracts and peduncles, and the tubular flowers are large for the size of the plant, pale rufus fading to pear-green with a strongly flared tip and noticeably hairy. In the open garden I have found this species to be more tender than *Beschorneria septentrionalis* and *B. yuccoides*, sustaining damage at –8°C from which it has taken several years to recover. With overhead cover cold is not such a problem: Tom Hart Dyke grows it well in his Mexican section at the World Garden in the frost pocket of Lullingstone, Kent, but with a temporary winter shelter. (H4)

When fact-checking *Beschorneria* species for this book, I encountered information that has thrown the identities of the above three into doubt. I am now 100 per cent certain that my purported *Beschorneria calcicola* is not that taxon. I also have a sneaky suspicion that the plant I described above as *Beschorneria tubiflora* is, in fact, *Beschorneria rigida*. This complicates things considerably as all the plants introduced into cultivation were grown from the seed of my plant as *Beschorneria tubiflora*, meaning a lot of people will need to change their labels. As things stand, this leaves 'my' *Beschorneria calcicola* and *Beschorneria rigida* stranded without correct identities.

Watch List

Beschorneria albiflora
(syn. *Beschorneria chiapensis*)

The odd one out in the genus, it is the only truly arborescent *Beschorneria*, growing trunks said to reach an

Beschorneria tubiflora.

Beschorneria tubiflora flower.

Beschorneria albiflora.

astonishing 8m, with secondary stems arising at the base. It typically has glossy green, leathery, 7cm-wide leaves to 70cm long with, unusually, smooth margins. Sometimes the leaves display a glaucous patina. The inflorescence is usually an erect, 2m-high, much-branched, rhubarb-red panicle with long, horizontal branches from which dangle flowers that are mainly pale iceberg lettuce green – not white, as implied by the name, but much paler than others in the genus. Its habitat ranges from the very south of Mexico to as far as Honduras and it is found in subtropical cloud forest growing in deep rich soil and shade. I have three clones of this, which all look rather different from each other. Two have flowered with different flower structures and colours, and the third, although it has not yet flowered, has completely different-shaped leaves reminiscent of sweetcorn. It would not surprise me in the least if future studies showed them to be different species. Two clones have been planted out in my garden for some years. With cover they are fine to –6°C: unprotected they are damaged at this temperature, but for milder, wetter climates than mine they represent a truly spectacular addition to the garden with their tall trunks and extravagant inflorescences.

Beschorneria albiflora flower.

A potted and flowering *Beschorneria albiflora* 'sweetcorn' form at RHS Wisley.

Glossy green, wavy leaves of the 'sweetcorn' *Beschorneria albiflora*.

Flower of *Beschorneria wrightii* in Mark Fillan's glasshouse, Devon.

Beschorneria wrightii

Having tried a few imposters over the years, I have only recently got hold of correctly identified material of this so, with me at least, it has little track record of being grown. My initial results show it to be either the least hardy or the most fickle of the genus as, outside, one plant died in the early cold winter and another has been damaged under unheated glass, so further trials are needed. It should have a very large and colourful inflorescence and, eventually, a small trunk with an affinity to the *Beschorneria yuccoides* group. Nurseryman Mark Fillan has raised a number of hybrids between this and other species that seem to be more robust.

Dasylirion

Family: Asparagaceae
Subfamily: Nolinoideae

Dasylirion is an architectural cousin of *Agave* and *Yucca*. Its distribution similarly centres in Mexico, to the north slipping over the border into some southern states of the USA and, to the south, coming to a halt roughly at the isthmus where the Mexican states of Oaxaca and Chiapas meet. It forms dense crowns of long, very narrow leaves that are mostly armed with prominent, marginal teeth. Some species grow trunks that can be sparsely branched, others remain trunkless with a branching, subterranean stem forming a slowly spreading clump. *Dasylirion* has separate male and female plants with, generally, showy, tall, erect and densely branched inflorescences of fairly insignificant flowers. A solitary female will still produce the puffy bladders that usually contain seed but, with no male, these remain empty. The period following the enormous effort of flowering is usually when plants are most weakened and they can enter a kind of hibernation, gathering themselves for anything up to a year before recommencing strong growth. *Dasylirion* will generally start again from a dormant bud to the side of the old flower spike; additionally it may (or may not!) offset from the rootstock underground.

Male flowers of *Dasylirion acrotrichum*.

Female flowers of *Dasylirion acrotrichum*.

There are something like twenty described species, most of which are locally called desert spoon or *sotol*, a handful of which are in cultivation and worth trying in UK gardens. One of the biggest problems is actually knowing which species you have. Plants from the commercial producers in Europe are of doubtful parentage: stock is usually raised by seed collected from field-grown, dioecious parents, meaning most are likely to be hybrids, whatever the label might say. Muddying the water further, there seems to be a kind of lazy default naming of plants: anything green with teeth is called *Dasylirion serratifolium*, anything with a blue tinge to the

Dasylirion glaucophyllum imposter at Naples Botanic Garden.

Dasylirions: one blue one, one green one.

leaves and teeth is called *Dasylirion glaucophyllum*, anything without teeth is *Dasylirion longissimum*. Ironically, true examples of all three of these species are actually extremely rare in cultivation. To be fair, for the majority of casual encounters with *Dasylirion*, that is probably good enough. As the famous American plantsman, Dave Ferguson, once said 'there are blue ones with teeth, green ones with teeth and a green one without teeth. Grow one of each and you've got dasylirions pretty much covered.'

Once you start looking more closely, however, there are interesting and ornamental differences to be seen, with each species bringing something unique to the party. Some have very attractively frayed leaf tips, forming a brush-type structure: others vary in the colour and size of the marginal teeth. Also, importantly, as the different species come from a range of climate zones, knowing exactly which you have makes potential garden hardiness less of a lottery. Thankfully plants grown from seeds of known provenance are becoming increasingly available.

Another Useful Plant

Sotol plants have been used since prehistoric times by indigenous populations for many purposes including food, drink, baskets woven from the leaves and building materials. As a source of food, the plants were dug up and roasted in a stone pit, the flesh then broken apart and eaten. It would have been mashed or, more usually, gnawed from the flared base of the leaf, leaving the unpalatable 'spoon' behind. Mixed with water and left to ferment, the mash turns into an alcoholic beer-like drink along the lines of *pulque*. After the Spanish conquest, an alcoholic spirit was made from them in much the same way as *mezcal* is made from agaves – the mash is distilled into the liquor simply called *sotol*.

Keystone Species

Dasylirion acrotrichum

Common name: sotol
Distribution: central Mexico
Hardiness: H4/5

Dasylirion acrotrichum (syn. *Dasylirion acrotriche*) has three recognised geographic varieties of which var. *acrotrichum* from central Mexico is the most ornamental, having been grown in many of the famous Italian and French Riviera gardens since the late nineteenth century. An iconic feature of the Tresco Abbey Gardens is the wonderful 'Neptune Steps' that are lined with venerable, sprawling plants of *Dasylirion acrotrichum*. Of the 'green ones with teeth' this one is probably the best garden plant, forming a densely compact dome of foliage to around 1.2m in diameter. Leaves are narrow, deep green with a matt surface and pale brown marginal teeth. Frayed leaf tips are a conspicuous and attractive feature of the plant. In time this can grow a tall trunk – in its habitat I have seen very old specimens up to 4m high – but it takes decades to get there. I have several of these in my garden and a couple of years ago two flowered at the same time: one was male,

Dasylirion acrotrichum, east central Mexico.

Dasylirion acrotrichum frayed leaf tip detail.

Dasylirion acrotrichum mother plant regaining her figure two years after flowering.

one was female and viable seed was set. Despite keeping the inflorescences in place for a long time, both plants have recovered quite quickly and two years on, the new crowns are full again.

Dasylirion acrotrichum flowering along the seafront at Southend, Essex.

I bought my first *Dasylirion acrotrichum* in 1996 and it has spent every single winter outside since then without any damage, flowering three times and developing a leaning 1m trunk with two new heads growing up from the base. It has immense character and is one of my favourite plants in the entire garden. Getting accurately named plants these days is a problem: *Dasylirion acrotrichum* is often sold as *Dasylirion serratifolium*. The likelihood is that anything labelled *Dasylirion serratifolium* is going to be one of these or, at least, a hybrid of it. (*See* the note on *Dasylirion serratifolium* later in this section)

Dasylirion quadrangulatum

Common name: Mexican grass tree
Distribution: north-eastern Mexico
Hardiness: H4/5

Dasylirion quadrangulatum (syn. *Dasylirion longissimum*) is a robust and tolerant species from north-east Mexico, where it is found over a large range in stands numbering countless millions of plants. This has very narrow, square-sectioned, dull green leaves held stiffly to form a beautiful shimmering 2m-wide globe atop a thick trunk that can reach 5m in a generation or two. Old specimens I have seen in their habitat have occasionally branched, making

Countless *Dasylirion quadrangulatum* extending to the horizon near Doctor Arroyo, Nuevo León.

Dasylirion quadrangulatum flowering in my front garden.

The scruffy silhouette of *Dasylirion longissimum* in habitat.

spectacular multi-headed plants. This is a singularly dramatic and architectural plant that deserves to be more widely grown if space allows. I have had two growing in my Essex garden for 25 years, with one more recent, larger addition, all of which have laughed at my worst three winters. It allows the gardener to grow a plant that has the same desirable form as the majestic Australian grass trees, *Xanthorrhoea*, but is much hardier and forgiving.

It is almost universally incorrectly called *Dasylirion longissimum*, a name formerly applied to all the toothless species. A study of the genus in the 1980s concluded that a population of plants in the centre of Mexico were different and it was these that were confusingly left with the species' name *longissimum* (via a couple of taxonomic red herrings). This is similar but scruffier and rare (but annoyingly not absent) in cultivation, leaving the overwhelming majority of toothless sotols to be *Dasylirion quadrangulatum*. To check for sure, rub your finger along the edge of the spoon (right at the base of the leaf where it widens). No residual spines and 'shoulders' on the edges of the spoon means that it is *Dasylirion quadrangulatum*. A few small residual spines means either *Dasylirion longissimum* or, as is often the case, a hybrid with one of the toothy ones, especially if the little teeth extend up the leaves a short way.

Worth Considering

Dasylirion berlandieri

This is another highly variable species with recognised geographic varieties that are mostly scruffy, non-descript grey-green things. The variety *berlandieri* from northern Mexico is an exception, making a large, open rosette some 1.5m in diameter with broad leaves of a strong pale blue colour. They are probably the most accommodating 'blue ones with teeth'

Dasylirion berlandieri in James Knock's garden, near Brentwood.

Dasylirion quandrangulatum close up of 'spoon' showing shoulders and no residual spines.

Dasylirion longissimum detail showing lack of shoulders on 'spoon' but residual spines.

Dasylirion quandrangulatum hybrid showing lack of shoulders and larger residual spines that extend further up the leaf margins.

and I imagine the chief suspect for most plants labelled '*Dasylirion glaucophyllum*'. Tom Hart Dyke has a plant growing in 'Mexico' at Lullingstone that has endured –17°C under the plastic cover. (H5)

Dasylirion miquihuanensis

This is a hefty brute of a plant with a massive 2m-crown spread and thick trunk to 3m or more high. Leaves are wide, stiff, olive green and have no conspicuous frayed tip. I have seen these growing on open, scree hillsides in the region of Miquihuana in north-east Mexico, where they grow into large, impressive plants in large, impressive colonies. Confusingly, there is another plant, as yet undescribed, that grows in almost the same area. Seed of this must have been collected and, because of its provenance, wrongly attributed to being *Dasylirion miquihuanensis*, and it can sometimes be seen for sale under this name. It is completely unlike *Dasylirion miquihuanensis*, being essentially similar to *Dasylirion acrotrichum* in that it is a compact plant forming a tight 1.2m-diameter dome of very narrow, mid-green leaves but with only a very short trunk. The terminal tuft on this form is exceptionally pronounced, initially twisted then dying off and fraying into a brush, and a very ornamental feature. Both can be found from specialist nurseries, so let your eyes, not the label, make your choice. Not widely grown but plants have survived –10°C. (H4/5)

Dasylirion miquihuanensis in Miquihuana, Tamaulipas.

Small *Dasylirion* sp. nov., also in Miquihuana, Tamaulipas.

Dasylirion wheeleri

Occurring across a wide geographic range throughout Arizona, New Mexico and the neighbouring north-western Mexico states, this grows in particularly arid regions and, despite extreme cold tolerance (–25°C), is less adapted to having wet feet than many other, less hardy dasylirions. Some forms of *Dasylirion wheeleri* can be quite blue, often they are dull green, and often plants will show a distinctive twist in the leaves. A group of three have been growing at RHS Hyde Hall in Essex for many years, one of which flowered recently. They are best grown in a position of full sun from dawn until dusk. (H5)

Dasylirion wheeleri in a front garden in Essex.

Watch List

Dasylirion glaucophyllum

A chunky and compact species from central Mexico with a dense crown to perhaps 1m diameter with fairly short, stiff, silvery-grey leaves that are very wide at the base. Eventually this will grow a short trunk. Having seen the true species in its habitat I am fairly sure that all plants I have seen labelled as this in Europe – including those in some venerable botanical collections, such as the plant growing in the Naples Botanic Garden – are not this taxon. As and when it enters cultivation it should, given its provenance, be a tough and hardy plant. (H4/5?)

Dasylirion glaucophyllum in habitat.

Dasylirion glaucophyllum grown from known provenance seed in a Suffolk garden.

Dasylirion lucidum in southern Mexico.

Dasylirion lucidum

This is a particularly pretty species from the Tehuacán valley along the Oaxaca/Puebla border in southern Mexico. It is a fairly compact species, with crowns reaching 1.5m across, eventually growing a trunk, offsetting at the base and even branching. Its main distinguishing features are glossy leaves (hence the name) and red marginal teeth. Plants of this species have only been available for the last few years and so it has not been widely tested in the UK. With such a southerly provenance you would not necessarily expect the plant to be very tough; however, my own plant has been through my early cold and late cold winters without protection and not been at all damaged. (H4?)

Dasylirion serratifolium

This is another instance where, having seen wild populations, I am not convinced that plants masquerading under this name in Europe are correct – more likely a catch-all name for all the mutts, hybrids and anonymous 'green ones with teeth'. The real species is a very robust plant with a 1.5m diameter, open crown of very stiff, wide leaves that have the texture of coarse sandpaper and which will eventually grow a thick trunk. It is one of the most southerly Mexican species from Oaxaca and southern Puebla and, as such, unlikely to be one of the hardier species. Certainly plants raised from accurately sourced seed died at just –5°C in my garden. (H3)

Dasylirion serratifolium in southern Mexico.

Furcraea

Family: Asparagaceae
Subfamily: Agavoideae

Furcraea are imposing close relatives of agaves and yuccas, with 26 or so species distributed from the centre of Mexico down through Central and South America across to the Caribbean. Many are large, trunked plants and they are often found growing at high altitudes. One feature most share is enormous branched inflorescences, which, in many species, are heavily laden with hundreds of bulbils. Some species reproduce only by seed, some will have a mix of both. The leaves can be a source of strong fibre that is known locally as *cabuyo*. There is a trunkless species, possibly *Furcraea hexapetala*, that has become feral in places such as Madeira, Portugal and the Canary Islands, and due to prolific bulbil production has become a real weed problem. Three of the Mexican species are worth growing outdoors in milder parts of the UK.

Keystone Species

Furcraea parmentieri

Common name: none applied
Distribution: central Mexico
Hardiness: H3/4

Furcraea parmentieri (syn. *Furcraea. longaeva*, *Furcraea bedinghausii*, *Furcraea roezlii*) is found on the high peaks of the volcanic belt across central Mexico, between 2,800m and 3,400m altitude and often above the tree line. It quickly grows an impressive solitary crown of glaucous grey, rough-textured, sword-shaped leaves that reach around 1.6m across, then goes on to form a 50cm-diameter trunk that, in its habitat, usually reaches around 4m. At maturity an immense inflorescence, which, at 8m high with multiple tiers of wide branches, is one of the biggest in the plant world, heads skywards. When fully developed the branches arch downwards with the weight of the bulbils produced along their length, a truly remarkable spectacle that I was once privileged to witness first hand on the peak of Mineral del Chico in central Mexico. Being monocarpic the plants then die, putting all their life's accumulated resources into this one final, phenomenal event.

Furcraea parmentieri is a fairly common sight in the UK's West Country, where they have been grown for well over a hundred years, with coastal towns in Cornwall and Devon usually boasting many of these monsters. In this country they never reach the same impressive size as in the wild, and plants will usually flower with maybe 1.5–2m of trunk. Confined in a pot they will often flower and die precociously at just a few years old, never managing to form a trunk. Not infrequently there is the remarkable spectacle of a '*Furcraea* Year' in which nearly all the plants in a particular area flower together: an astonishing event and one that leads to a glut of bulbils by the end of the season. It is safe to assume that this mass flowering is because they were all grown from bulbils produced during

Furcraea parmentieri at 3,000m altitude on the high peak El Chico, Hidalgo.

A massed flowering event at Tremenheere Sculpture Gardens, south-west Cornwall.

Impressive inflorescence of *Furcraea parmentieri* at Tremenheere.

Furcraea parmentieri not so happy in my dry Essex garden.

Furcraea parmentieri towering above me in the balmy Madeira climate.

the last such year, and thus they are the same age.

What actually triggers flowering is open to conjecture: it has been postulated that a spell of unusually cold weather or other form of physical stress, such as being pot bound, might hold the key. Certainly there is a group of three *Furcraea parmentieri* growing near the toilets at the Jardines Palheiro on the outskirts of Funchal in Madeira that enjoy a stress-free existence in the especially temperate climate there. As yet they have shown no inclination to start flowering and simply keep growing taller and taller. Although, interestingly, the last time I was there they had started to produce bulbils directly on their trunks.

One of the more vexing aspects about *Furcraea parmentieri* is sifting through all the synonyms and working out exactly to which plants people are referring. Historically in Europe this species has always been called *Furcraea longaeva*. A revision of the arborescent Mexican *Furcraea* changed this, utilising the name *Furcraea longaeva* for a southern species, not the central Mexican one, but it is still the name you mostly see for this species. It is still common in literature to see references using the names *Furcraea roezlii* and *Furcraea bedinghausii*.

I struggle to grow this well in my garden. It prefers the cooler, wetter and milder climate of the south-west and my plants are usually damaged irreparably by exposure to –6°C for anything but the briefest time. With a cover they tolerate cold better but then suffer for being too dry, as they enjoy regular rainfall throughout the year. Where the climate allows, however, this is hands down one of the most spectacular plants it is possible to grow in a UK garden. It is worth all the effort to try and secure the sensational flowering.

Watch List

Furcraea longaeva

The 'real' *Furcraea longaeva* is a huge plant growing in Puebla and Oaxaca in the south of Mexico. Compared to *Furcraea parmentieri*, the leaves are longer, thicker and deeply guttered, the crown larger and more robust. Plants of this species have not been in cultivation in Europe for long but if you are in the milder south-west, I would say they are definitely worth seeking out. I have a plant that has taken –4°C with no damage but been quite badly damaged by my early cold winter. West of Penzance this would be stunning. (H3)

Furcraea quicheensis

A more compact species from Chiapas in the very south of Mexico and into Guatemala and Honduras, compared with *Furcraea parmentieri* this has a dense crown of shorter, broader, softer, grey-blue leaves, a shorter trunk and, should one flower, it is a much smaller affair. I would say that as a young plant it is more attractive as the chunkier leaves just seem better in proportion. This one has even less of a track record as I have grown one in my garden for just two full growing seasons so far. It has grown very quickly into an extremely handsome plant, not yet trunking, and taking a very mild winter with no damage. My early cold winter, however, caused it a fair bit of damage, from which it is now recovering. Again, for growers in the milder, wetter parts of the UK this is worth seeking out as it is a bit different. (H3)

Furcraea quicheensis.

Furcraea longaeva in southern Mexico.

Hesperaloe

Family: Asparagaceae
Subfamily: Agavoideae

Hesperaloe is a genus of eight species of xeric plants closely related to *Agave* and *Yucca* from Texas and north-east Mexico. They are evergreen perennials with long but narrow, guttered leaves, usually with curling fibres along the margins, similar in appearance to some of the stemless *Yucca* species, growing into loose, open rosettes and, in time, dense clumps. Containing strong fibres, the leaves have been utilised for making ropes, baskets, textiles, etc. for hundreds of years. These are popular xeriscape plants in the USA and Mexico being extremely tolerant of drought and heat. Only a handful has made it into cultivation in Europe.

Keystone Species

Hesperaloe parviflora

Common name: red yucca
Distribution: north-east Mexico, Texas
Hardiness: H5

Of all the *Hesperaloe* species, this has the longest track record for being grown outside in the UK. It is slightly more compact than many others in the genus, comprising rosettes of dark green, narrow 80cm-long leaves with pronounced white, curling fibres. Plants will often flower when young, producing their showy, tight 1.5m-high panicles of flared tubular flowers if grown in a sunny position. Flower colour can vary from pink, through red into yellow via orange, most commonly coral pink on the outside with an orange-red interior. The plant will offset and thus eventually grow into a clump of many crowns. In habitat it is found growing in harsh conditions, rocky limestone slopes that experience extended drought, high summer temperatures and extreme cold, down to –25°C in some locations.

In the UK it is best grown in the hottest and sunniest spot you can give it. I have found it hard to kill but even more difficult to grow well. My own original plant languishes somewhat in partial shade but subsequent plants have performed better, even flowering, in positions of full sun. They have all taken my coldest winters with no issues. This species is fairly widely available through exotic plant nurseries in the UK and is well worth growing. Even without flowers, the rosettes of dark green leaves with their contrasting white filaments are attractive. Flowers, when they occur, are just a spectacular flash of colour.

Hesperaloe parviflora in an Algarve garden.

Hesperaloe parviflora flower detail.

Watch List

Hesperaloe funifera

This is the largest species in the genus from central and north-eastern Mexico: I have seen this growing in dry scrub at around 1,800m in Tamaulipas near Jaumave on a bumpy road down to 'El Cañón del Salto'. Most of the plants were buried in a thorny acacia of some sort. Leaves are yellowish-green, up to 1.8m long and held stiffly erect. The flower spike is a huge 3.5m or more long, sparsely but widely branched affair with creamy white starry flowers. I grew this many years ago, when it slowly made a tight clump from a single rosette over a period of five or six years. Sadly, before it ever got to flowering I lost it in one particularly wet summer. I was keen to replace it and, indeed, thought I had until the plant flowered and it became clear it was *Hesperaloe nocturna*. I am still looking. In habitat these are very hardy, taking –15°C. (H5)

Hesperaloe funifera in the Desert Botanic Garden, Arizona.

Hesperaloe malacophylla

Only described in 2010, this is an interesting species as it grows in partial shade, in forested areas in the Mexican state of Tamaulipas, in relatively moist conditions. It is similar in size to *Hesperaloe parviflora* but paler green with less pronounced marginal fibres. I have actually been growing this since just before it was formally described, acquiring a couple of seed-raised plants as sp. nov. that have grown slowly but steadily in that time. One in full sun has performed better than the one in shade, despite the forest provenance. I would say this shows great promise being, in theory, more suited to UK growing conditions. My plants have

Hesperaloe malacophylla.

taken my early and late cold winters in their stride with no damage. (H4/5?)

Hesperaloe nocturna

I got this entirely by mistake believing that I had bought a replacement for my *Hesperaloe funifera*. It seemed, at the time, very similar, maybe a little smaller, with less of a yellow tinge in the leaf. I assumed it was just a slightly different form but the flowers gave it away – the pale greenish-white stars, with a hint of pale mauve – opened at night, which confirmed the identity. For me this was very exciting as I had not seen *Hesperaloe nocturna* offered for sale anywhere, so to actually get hold of it accidentally was serendipitous! This has taken my early and late cold winters without issue and, as it has flowered, I think maybe it is less heat-demanding and more tolerant than *Hesperaloe funifera*. (H4/5?)

Hesperoyucca

Family: Asparagaceae
Subfamily: Agavoideae

A genus of two species recently split from *Yucca* that are found in Arizona, California and down into Baja California. They are trunkless and monocarpic, taking many years to reach maturity but providing a jaw-dropping spectacle when in flower. The plants were used historically by indigenous Americans as a source of food – the heart of a plant about to flower cooked in a pit – and fibre from the leaves employed for a host of uses once processed. For such a small genus it seems to have had more than its fair share of taxonomy debates. Some authorities recognise three species, with anything up to six subspecies for one of them, whilst some authorities consider these to be just varieties.

Keystone Species

Hesperoyucca whipplei

Common name: Our Lord's candle, chaparral yucca
Distribution: north-east Mexico, Texas
Hardiness: H4

Hesperoyucca whipplei (syn. *Yucca whipplei* and including subsp. *caespitosa*, *intermedia*, *percusa*, *eremica*, *parishii*) grows into a 1.5m-diameter, tightly packed dome full of seriously dangerous leaves. These are short, narrow, slightly guttered, sharp-tipped with finely serrated margins in the most gorgeous pale china-blue colour. At maturity, around ten years in their habitat but a bit more in cultivation in this country, the plant throws up the most incredible 3m-high flower spike, clothed in the upper half with short side branches upon which are carried thousands of creamy white, bell-shaped flowers. In their habitat, California down to Baja California, the plants will take down to –12°C.

These have been grown outside in the UK in a number of locations for many years, always hitting the headlines when in flower. Kew Gardens had one in flower a few years ago near the Alpine House and, more recently, Tom Hart Dyke in the World Garden at Lullingstone. Tom's plant was kept covered each winter, which allowed it to reach maturity. I have yet to grow one to flowering size – my latest attempt, having got off to a slow start, is gradually picking up momentum.

Tom Hart Dyke admiring his flowering *Hesperoyucca* at The World Garden, Lullingstone, showing his usual reserve.

If you have, or can get, one of these beauties it is definitely worth growing. In leaf it has great character, a prickly, tightly packed dome of leaves, but in flower it is breathtaking.

Manfreda

Manfreda was a genus of plants related to *Agave* but recent DNA studies have shown that they are not really that distinct after all and so they have now been placed into their own group within the genus *Agave*. Gardeners disagree, as usual, as manfredas look and behave completely differently to agaves – being largely deciduous perennials with long, spidery, tubular flowers – but it seems we are stuck with it. As *Manfreda* or *Agave* I have never found any of them to be particularly garden worthy. Many years ago I grew what was *Manfreda maculosa* (now *Agave maculata*) but snails ate it more quickly than I could keep pace with. I have tried *Agave guttata* but lost it to –4°C. Where the interest in the genus lay more recently is when it is hybridised with *Agave* to give a race of ornamental plants known as × *Mangave*, with increasingly large numbers of named cultivars and hybrids appearing on the market with increasingly catchy names. Adopting the new genus change, how-

The unfeasibly colourful × *Mangave* 'Espresso' in Chris Grundy's garden.

ever, means that these should no longer be referred to as × *Mangave*, just complex *Agave* hybrids. They are primarily bred for colour – spots, variegation and shading – and bizarre growth forms, and have become incredibly popular as potted plants, widely distributed via garden centres and even supermarket chains. They are not really intended for permanent planting outdoors, certainly as far as the UK is concerned. One or two have been tried outside historically, the most successful to date being one of the earliest cultivars *Agave* 'Bloodspot', which, if kept dry, can take –5°C or lower. I recently tried the cultivars 'Red Wing', 'Blazing Saddles' and 'Pineapple Express', which all struggled back rather weakly in mid-summer having been defoliated at –5°C. With so many potentially hardy *Agave* species I have yet to try, I have little enthusiasm to try any more mangaves.

Nolina

Family: Asparagaceae
Subfamily: Nolinoideae

A genus of succulent plants distributed throughout Mexico and into the southern USA; historically, but not currently, including plants of the genus *Beaucarnea*. There are some 30-odd species, all dioecious, split between tall, imposing arborescent plants and low grassy mounds sprouting from an underground caudex. Take a few steps backwards and many species can appear superficially similar – especially the trunkless, grassy species. But get in close and the detail sets each apart. Most have leaves that are minutely serrated on the edges, which can cut flesh easily and painfully, so handle with care. Often the trunked species will expand at the base showing a small 'elephant's foot'. All can be readily propagated from seed. Horticulturally they are criminally overlooked and their potential in the xeric garden is, as yet, untapped. In the USA they are almost universally called 'beargrass' with a variety of indigenous names also being used.

Keystone Species

Nolina hibernica

Common name: mountain beargrass, nolina 'la siberica'
Distribution: mountains of north-east Mexico at elevations up to 3,200m
Hardiness: H5/6

Nolina hibernica is a superlative specimen plant for the xeric garden, becoming more interesting and dramatic in appearance with age. Arching leaves, up to

Nolina hibernica in its woodland habitat at 3,000m altitude on Cerro Peña Nevada, Tamaulipas.

100cm in length and 5cm wide, billow and swirl from a central crown, usually with a twist at the tip. It can quickly grow a stout trunk that is attractively fissured with corky wings; it soon creates the perfect pedestal for the leaves. If allowed to remain in place, once they have died off the leaves form a parchment-like twisted petticoat that is, in itself, a really attractive feature. Plants can flower relatively prematurely, thrusting skywards a huge, frothy plume of hundreds of tiny little creamy white stars, after which it is common for growth to resume from multiple crowns formed around the base of the old flowering stem, although most often settling down to two or three, aborting the rest.

Perfect, twisty-leafed specimen of *Nolina hibernica* at the World garden, Lullingstone.

Nolina hibernica showing swirling eruption of leaves from the crown.

First described from the small village of Siberia on the high mountains bordering the Mexican states of Nuevo León and Tamaulipas, and indeed originally known by the working title *Nolina* 'La Siberica' until formally described a few years later, this can be found growing on both open hillsides and as understorey in pine/oak woodland alongside plants such as *Agave montana*. The climate in these mountains is cool in summer, cold, wet and misty in winter with frequent snowfalls, making it the perfect arborescent *Nolina* for cooler, damper temperate climates.

In cultivation *Nolina hibernica* is proving to be easy to grow, tolerating a range of soils, as long as they are free-draining, and able to cope with a wetter climate than most succulent plants. It is certainly extremely hardy, documented as having endured temperatures as low as –18°C without ill-effect in the centre of the UK. Once settled in, it is perfectly reasonable to expect the growth rate to reach 10cm a year, especially if given a position in full sun.

Nolina nelsonii

Common name: blue beargrass, *Soyate*
Distribution: Mexico – restricted to a small area in the north-western part of Tamaulipas State
Hardiness: H4/5

Nolina nelsonii is a plant with impact at all sizes, superficially resembling an outsized yucca and performing the same role in the garden. It has a dense crown of rigid, steely blue-grey, sword-like leaves, 50–90cm long and 2–4cm wide, held stiffly in a globe as if the product of an artist working in stainless steel. Unlike yuccas there are no sharp terminal spines but the tips can dry off leaving them brown and withered. Older plants will form a trunk perhaps reaching, after a generation or two, 4m in height and 25–30cm across, but making steady progress in the meanwhile. These old trunks are sheathed in a petticoat of straw-coloured dead leaves. Older plants – certainly trunked plants – will flower, producing an extravagant, broad panicle of creamy white flowers between 2m and 3m long – after which some plants may branch, while others remain single-headed. In the field, the most branching I have seen gave four heads, though a particularly robust cultivated plant in a garden nearby had managed seven.

In habitat the plant is visible as bright silver-blue flashes punctuating arid limestone scree hillsides to around 2,200m altitude, growing alongside plants such as *Brahea decumbens*, *Dasylirion quadrangulatum*, *Dasylirion miquihuanensis*, *Agave americana* subsp. *protamericana*, *Agave striata* and various small pine and juniper species.

In cultivation *Nolina nelsonii* seems perfectly accommodating and less fussy about soil conditions than some, as long as it is well drained. Large plants have been known to survive –12°C in the open garden in the UK, lower if covered. Growth rate is steady – once at maximum width they can grow around 10cm of trunk a year – more if fed and watered. Given their liking for open hillsides in habitat, I would give *Nolina nelsonii* plenty of room in full sun.

These spectacular plants are becoming increasingly available in the UK and are highly recommended, being every bit as dramatic as some of the more well-known trunked yuccas, such as *Yucca rostrata*, with, ultimately, more presence.

Nolina nelsonii in habitat near Miquihuana, Tamaulipas.

An eye-catching specimen of *Nolina nelsonii*.

Nolina parviflora.

Nolina nelsonii flowering at Ventnor Botanic Garden, Isle of Wight.

Nolina parviflora

Common name: Mexican grass tree
Distribution: most Mexican states
Hardiness: H3/4

Nolina parviflora (syn. *Nolina longifolia*) is an extremely common plant in Mexico with an extensive distribution in the more central, highland states from Chihuahua in the north all the way down to Oaxaca in the south. Different populations can vary considerably in appearance but, typically, they form freely branching, slim-trunked trees, to around 6m tall, each stem topped with an erect tuft of narrow, downward-curving leaves that, once dead, are retained on the trunk as a 'petticoat'. Plants grow multiple heads after flowering – itself a memorable event as it produces a 1.5m-high dense plume of tiny cream stars. The most impressive forms I have seen were growing in the high mountains of western Querétaro: massive plants with 10m trunks, dense crowns of very stiff leaves and a particularly thick, shaggy petticoat of dead leaves, reminiscent of *Nolina hibernica* but a good deal taller. Sadly, these have yet to make it into cultivation.

An extremely robust form of *Nolina parviflora* in Querétaro.

Plants from the southern end of the range in Oaxaca and Puebla were previously known as *Nolina longifolia* but these are now considered to be *Nolina parviflora*.

A distinct form of *Nolina parviflora* with exceptionally long leaves.

Second specimen of *Nolina parviflora* showing a colourful flower bud.

Around 25 years ago I obtained some plants labelled as *Nolina longifolia* that stood out as being rather different from most and which have grown into the most gorgeous plants. The leaves of this nolina are extremely long, sometimes over 2m in length, and 5cm wide at the base. As they lack the stiffness seen in many other nolina leaves, they immediately droop downwards, which, on young plants, can look a bit untidy (although perfect for a tall pot). As my plants have aged they have developed stunning thick trunks, swollen at the base and covered with markedly protuberant corky ridges (more pronounced than those of other nolinas I grow), allowing the weeping crown to cascade down the length of the plant. My plants have survived my three worst win-

Trunk detail of the same plant showing pronounced corky ridges.

The nursery trade, however, blithely perpetuates the name *Nolina longifolia* (or even *Dasylirion longifolium*) for nearly all trunked *Nolina* species, regardless of provenance. Having such a large range in the wild means individual populations can experience very different growing conditions, from misty cloud forests to arid, open plains. Without any knowledge of provenance it is, therefore, impossible to predict the cold-tolerance of plants sold simply as *Nolina longifolia*: I have experienced ostensibly similar-looking plants unconcerned by –8°C, when others were killed.

Same specimen in full flower.

ters with minor foliage burning, although I suspect these were close to their limit. Venerable specimens of this pendant form can be seen in many of the older European arid collections and gardens, where the swollen elephant foot base gives rise to many stems, sometimes over 3m high.

All forms of *Nolina parviflora* are worth growing, especially if you hit upon a cold-hardy one but, if your climate allows, this long-leafed form really is a spectacular focal point for the xeric garden – the unique silhouette, weeping foliage and distinctive corky stems setting it apart as major talking points.

Nolina texana in habitat.

Nolina texana

Common name: Texas *Sacahuista*
Distribution: southern states of the USA down into northern Mexico
Hardiness: H6

Nolina texana is a stemless, evergreen succulent plant that grows from a slowly spreading underground, branching caudex. It forms a dense mound of tough, narrow, grassy leaves up to 120cm in length but only 4–5mm in width that are traditionally used in basket-making. Tips of the leaves will wither and dry into a curious 'shepherd's crook' shape, adding interest. Often larger plants can appear to look flattened as the weight of the accumulated leaves presses down the whole plant, lending an attractive weeping habit. In spring or early summer each mature crown can send out a densely branched short inflorescence that stays nestled within the leaves. Flowers are mauve in bud, opening to creamy stars in a narrowly conical panicle.

Typically seen in a habitat dotted around scrubby arid grasslands, rocky plains or nestling between boulders the plants are easily mistaken for grasses. In fact, historically, there was an ancient plant outside the south end of the Princess of Wales Conservatory at Kew Gardens, London, growing quite happily alongside two pampas grasses and masquerading as such, until giving the game away by flowering.

Nolina texana is unaccountably rare in cultivation but a perfect foil for the showier impact plants. Impervious to any degree of drought or cold it is likely to face in the UK, all it asks is perfect drainage. Given a position in full sun – or even part shade – it is worth considering as indestructible, large-scale ground cover. An annual grooming of the oldest leaves, once they have become straw-like, keeps the whole plant looking tip-top.

Nolina texana inflorescence detail.

Nolina texana close up shot of male flower.

Worth Considering

Nolina beldingii

Hailing from Baja California, this is potentially one of the tallest of the trunking *Nolina* species with long flexible leaves that end with a withered shepherd's crook tip. This is fast-growing and hardy to at least –15°C. (H4/5)

Nolina durangensis

The plants circulating under this name may or may not be this taxon but grow into handsome plants with beautifully arching leaves and slowly develop a trunk. Hardy to at least –10°C. (H4)

Nolina erumpens

From Texas, New Mexico and northern Mexico, this plant is slightly larger than *Nolina microcarpa* with a more prominent whiter inflorescence.

Nolina greenei

From New Mexico and Colorado, this is similar to *Nolina texana* but smaller, with wider leaves and a more prominent inflorescence.

Nolina lindheimeriana

Native to Texas, this differs from *Nolina texana* by being smaller, but with a much showier, more open inflorescence of creamy flowers held above the foliage.

Nolina microcarpa

From Arizona and New Mexico, is similar to *Nolina greenei* but with more rigid leaves and whiter flowers in an inflorescence held on longer stems above the rosette.

All these trunkless nolina are super-hardy. (H5/6)

Nolina lindheimeriana at East Ruston Old Vicarage.

Watch List

Nolina excelsa

Only recently described, this gauntly elegant, trunking nolina is from southern Mexico. It is almost like a bridge to *Beaucarnea* with a very pronounced swollen foot, deeply ridged corky stem and short, stiff, glaucous leaves. A small plant was defoliated in my early cold winter but has since sprouted from the stem. (H3/4?)

Nolina matapensis

Potentially one of the most ornamental of all arborescent nolinas, with fine, weeping, glaucous grey foliage and showy greenish-white plumes of flowers. From northern Mexico, it is not widely grown but has taken –6°C in my garden as a young plant. (H4/5?)

Nolina matapensis in Tenerife.

Nolina parryi

From Baja California across to Arizona, this must be one of the jewels in the nolina crown. A massive plant as it ages with stiff, broad leaves and an immense trunk up to 60cm across. Slow from seed. As a very young plant this has taken my early cold winter with no damage. An exciting plant. (H4/5?)

Yucca

Family: Asparagaceae
Subfamily: Agavoideae

Yucca is a genus of around 50 species of succulent, evergreen, rosette-forming trees and shrubs from much of continental North America, as far north as Alberta in Canada, down to Guatemala in Central America and over to some of the Caribbean Islands. Some grow into many-branched, towering giants on stout trunks, others remain at ground level and grow into multi-headed clumps but, as a rule, they all produce showy, prominent plumes of flowers. Uniquely in the woody lily world, moths are the sole pollinators of yucca flowers, sometimes with species-specific relationships, so outside of their natural habitat, with the exception of *Yucca aloifolia*, they are rarely seen carrying seed pods. When formed, seed pods are of two types: those that dry, split and scatter seeds like pepper pots when shaken by the wind; and thick, fleshy fruits that are dispersed by animals when eating the fallen sweet fruit. They are peerless as architectural specimens for the dry garden and represent the largest and hardiest components for UK xeriscaping, many of them apparently completely at home in our cool, damp winters and oblivious to freezing temperatures, ice and snow.

Keystone Species

Yucca aloifolia

Common name: Spanish bayonet, aloe yucca
Distribution: Atlantic and Gulf coastal states of the USA, parts of the Caribbean
Hardiness: H5/6

Yucca aloifolia is a strikingly ornamental shrub that forms erect stems, often clustering at the base, to 3m. Leaves are variable in length, some forms being rather compact at 30cm, others twice that length, arranged around the stems looking like a dense and rather scary bottle-brush. They are narrow and deeply guttered with finely serrated margins and sharply spined tips. The stems are relatively thin and flexible. Flowering shoots are sturdy panicles to 1.5m high, festooned with creamy-white open bells, often tinged with purple on the reverse.

This species, described in the early 1600s, has been known longer than any other yucca. It is commonly grown in many parts of the world, including the UK, where, of the 'exotic' species, it has the longest record of cultivation. It is extremely hardy, taking down to −18°C in its habitat – in my garden, plants took my worst winter without flinching. It is not so fussy about the soil in which it grows, as long as it drains well. There are three things to be wary of, however. First, in a world of spiky plants this is exceptionally spiky: the form of the leaves gives them great structural rigidity and they do not bend if you brush against them. The relatively slender stems can also flex away at first contact, then come back at you with those unyielding spine tips. The second thing

Yucca aloifolia, far too close to my front porch.

A slightly shorter-leafed form of *Yucca aloifolia* tucked out of harm's way.

Yucca aloifolia showing characteristic bumps on leaf bases and serrated leaf margins.

to be wary of is that they grow very easily from root suckers, not a problem until you move one and it keeps coming back from whatever tiny scraps of root are left in the soil. I first planted mine near the front porch, which, with hindsight, was not a great idea given the spikiness. Fifteen years later I am still removing suckers. Most importantly all parts of the plant are poisonous, especially the roots and especially to dogs.

Sometimes plants named *Yucca aloifolia* 'Purpurea' are available in nurseries with purple-tinged, recurved leaves but these are, in fact, a different and hard-to-please species, *Yucca desmetiana*.

Yucca faxoniana

Common name: faxon yucca, giant dagger
Distribution: northern Mexico, New Mexico, Texas.
Hardiness: H5/6

Yucca faxoniana (syn. *Samuela faxoniana*, *Yucca australis*) is a large, dramatic, chunky plant with one of the fattest trunks of any yucca – up to 1.5m wide on old specimens. They grow solitary stems up to 6m tall with stout, inflexible and fairly wide, yellow-green leaves to 1m long and 8cm wide. These have a brown, horny margin with, to varying degrees, thick curling white filaments. Flower spikes are jaw-droppingly spectacular – a 1.5m-tall, 1.2m-wide, many-branched panicle that is held above the leaves and dripping with hundreds of white bell flowers.

Its range is concentrated to either side of the border around the Big Bend National Park, growing in vast numbers on dust-dry plains where winters will see temperatures plunge to –18°C. Yet, despite this apparently harsh environment, plants seem to be quite happy growing in the UK given good drainage and a position in full sun. I have had a lovely specimen planted in my garden for nearly twenty years, which has never been affected by weather and has flowered just once. I planted it at the top of a steep bank and, thinking I was being clever, angled the stem rather than

The jaw-droppingly spectacular inflorescence of *Yucca faxoniana*.

My beefy specimen of *Yucca faxoniana*, propped up with an Acrow.

Detail of *Yucca faxoniana* curled filaments at leaf tip.

Yucca faxoniana showing accumulation of leaf fibres near the crown.

having it bolt upright. The weight of the plant slowly but surely dragged the plant downwards, to the extent that it now needs to be propped up by an Acrow support. It certainly looks interesting, anyway. *Yucca faxoniana* would make my top ten 'must grow' plants for a xeric garden in the UK, as nothing comes quite as close for sheer brute presence.

There is a similar species from further south in Mexico, *Yucca carnerosana*, that some authorities believe is just a geographic variant of *Yucca faxoniana*. Plants of this taxon that come into cultivation tend to be a little smaller, with shorter leaves carrying much more noticeable marginal filaments. They are also said to be less hardy but I have not seen hard data to back this up.

Yucca filifera

Common name: *Palma China*, *Izote*
Distribution: north-eastern to central Mexico
Hardiness: H5

Yucca filifera is a stunningly attractive plant at all stages of growth. Young plants soon grow that strange spiky bottlebrush look; middle-aged plants tend to remain a solitary trunk for decades, getting taller and taller before branching. Ancient plants are simply spectacular: huge, many-branched, 10m-high giants with immense character. Leaves are olive-green, narrow but almost uniform in width, to 50cm long and armed with a

Millions of *Yucca filifera* in habitat on a plain extending as far as the eye can see.

vicious spine at the tip. The leaf margins generally have many whitish, curled leaf filaments to a greater or lesser degree. Unusually, the inflorescence of creamy white flowers hangs downwards rather than being held erect; a reasonably good diagnostic clue when looking at these in the wild where their range overlaps similar-looking species.

These grow in huge numbers, covering enormous areas from central Mexico to the northern States of Coahuila and Nuevo León. Driving north from Mexico City along Highway 57, once you have passed the urban sprawl, the industry and finally the agriculture, you start driving through vast largely untouched plains and it is here where you see countless billions of these plants. For hour upon hour as you drive along the motorway all you can see are forests of these yuccas, stretching off into the distance for as far as the eye can see in all directions. It was my first daylight encounter with Mexican flora and one I have never forgotten.

Yucca filifera 25 years from planting – slow but steady.

Crown detail of *Yucca filifera* showing leaf filaments.

I planted my first specimen of *Yucca filifera* around 25 years ago as a 60cm tall plant and it has slowly but steadily grown into a 2.5m-high plant, shrugging off my worst winters with no sign of damage. I have kept all the old leaves in place: as they age and turn brown they turn sharply downwards, overlapping to create a protective skirt around the lower part of the trunk. I believe this goes some way towards protecting the plant against the ravages of winter as I have lost plants, bought as larger specimens that had been trimmed of their old leaves so that the exposed trunks rotted. Under cover and kept dry these are documented to take down to –17°C in the UK.

Yucca gloriosa

Common name: Adam's needle, Spanish dagger
Distribution: south-eastern USA
Hardiness: H6

Yucca gloriosa has been grown in Europe and the UK since the eighteenth century and is one of the most easily grown of the genus, tolerating a wide range of soils and conditions. I have even seen a picture of one growing with permanently wet feet at the side of a river. In recent years the biggest problem seems to be getting accurately named material. Mostly the plants sold under this name are inferior hybrids of some sort and generally grow into scruffy, grey plants that sprawl all over the place. The 'real' *Yucca gloriosa* is quite distinct with dark green, sometimes glaucous, broad, lance-shaped leaves, sometimes slightly pleated. These leaves are rigid and sharp-tipped, growing to 50cm, arranged in compact whorls to form dense rosettes. Plants grow a short, stocky, fibrous trunk, to maybe 1.5m, which is usually much branched at the base. New stems arise from the rootstock (growing points presenting as knobbly bumps often referred to as 'toes') and the whole plant eventually forms a multi-headed and rather formidable colony that can extend to 3m across. Creamy white flowers, often with a purple reverse, are borne on a 2m-high panicle produced late in the year, sometimes not fully opening before the first frosts spoil the party.

These are coastal plants, in nature growing in sand dunes along the south-eastern states of the USA from the bottom of Virginia to the top of Florida. In habitat and cultivation these are extremely hardy, taking temperatures down to –20°C and experiencing all types of adverse weather conditions.

I have admired this in many gardens since first becoming interested in plants. It is common to see a venerable old mother plant in a front garden somewhere and increasingly smaller plants in neighbouring plots – no doubt 'gifts' taken from the mother as it has become

Yucca gloriosa along the seafront at Southend-on-Sea.

Closer look at the uncompromising crown of *Yucca gloriosa*.

larger and larger. There are some particularly fine old plants in the promenade gardens along the seafront at Southend-on-Sea in coastal Essex that I imagine date back to Victorian times. Yet I have only recently acquired one for myself, as a recent garden expansion allowed me the luxury of a few plants that will get large in time. It was a gift from a friend whose 'mother' plant was starting to obstruct his path.

There is one recognised variety, *Yucca gloriosa* var. *tristis*, that was formerly known as *Yucca recurvifolia*. This is another splendid plant with a softer, less intimidating appearance and tolerant of a slightly shady position – always a valuable asset for succulent plants. It is inclined to grow a slightly taller and neater trunk and not so inclined to sprout from below ground. I used to regularly drive past a really old specimen of this, long gone now, that had grown into a many-branched tree over 4m high. Flowers are similar to the type but produced earlier in the year.

The more benign *Yucca gloriosa* var. *tristis* in a Cornish garden.

Yucca linearifolia

Common name: none applied
Distribution: north-eastern Mexico
Hardiness: H5/6

Yucca linearifolia has it all: looks, personality, easy culture and hardy to cold. Plants form a dense head of very narrow, stiff, four-angled to flattened, blue-green leaves that are held in a perfect globe to around 90cm across on top of a trunk that can be anything up to 3.5m tall. Plants offset sparingly at the base, especially when older plants collapse, forming small groups of trunks with staggered heights. Creamy white flowers are held in a showy conical panicle above the leaves. If not removed, the dead leaves bleach to a parchment colour and are held tightly to the trunk, forming a full-length petticoat that only adds to the beauty of the plant.

Only described as recently as 1995, this species has a fairly wide scattering of populations in the states of Coahuila and Nuevo León, Mexico. They are often seen on the north side of a hilltop ridge or shady side of a canyon, sometimes in apparently moist soil, as well as in more open situations. Populations vary in colour and size: plants you see on the road between Rayones and Galeana are quite blue with wide, robust crowns, whereas other populations, such as those at General Cepeda, are smaller and greener. It clearly has a strong affinity with the very similar *Yucca queretaroensis*, although the latter is not quite as hardy or accommodating.

In cultivation these are extremely tolerant of UK growing conditions, being happy with a slightly moister soil than most yuccas. They are extremely cold-hardy: a friend in the middle of Belgium has been growing these in his garden for many years, seeing down to –22°C. They do not seem to resent root disturbance and will re-root quite happily if transplanted – something that plants like *Yucca rostrata* or *Yucca thompsoniana* can struggle with quite often. I have heard of one occasion when a trunk rotted in the centre. The top was cleaned back to sound wood, re-rooted, then grown again in its shorter size. The base, meanwhile, sprouted offsets and regrew into a clump.

Yucca hybrid of indeterminate origin, still doing a good job.

Yucca linearifolia flowering spectacularly.

Yucca linearifolia in habitat in north-east Mexico clinging to a rock face.

For some reason it is unusual for this species to set seed, even in its habitat. One time when I stayed in Saltillo, Coahuila, the hotel had extensive plantings of these beautiful plants in the grounds. Thinking I might have stumbled upon a potential source of seeds for the future, I asked the proprietor how often they set seed. She, in turn, asked her gardeners, who replied 'never', as they had always propagated the plants using the *hijos* – children – which is how they referred to the offsets. I have recently discovered, however, that it is possible to achieve a heavy set of seeds if the flowers are hand-pollinated assiduously morning and evening while open.

For me this is easily the pick of all the exotic *Yucca* species that are available and one of my top three xeriscape plant choices. From small plants that have not yet made a trunk to large impressive specimens, they are always eye-catching. If you have space for just one yucca, ignore the others and seek this out. If you have room for several yuccas, get lots of these!

Yucca rostrata

Common name: beaked yucca
Distribution: northern Mexico into western Texas.
Hardiness: H5/6

Yucca rostrata has been grown in Europe for over a hundred years. Well grown, it is a stunning plant with a dense crown of pale blue-green leaves, narrow but flattened with a pale margin, on slender trunks

A perfect *Yucca rostrata* in an Italian garden.

An unexpectedly perfect *Yucca rostrata* in Kris Swaine's wet Wakefield garden.

to 4m high, often more of a bottle brush than a globe and keeping its leaves down the stem in bottle-brush fashion. Sometimes trunks will produce branches but not many. A huge plume of creamy white flowers is held above the leaves. Trunked plants are often available, usually trimmed of their older leaves so that the remaining stubs show interesting patterns.

This is a species from open, hot and very dry habitats, taking temperatures as low as –15°C with no issues. In the wild, different populations vary markedly in appearance, from tall, solitary-trunked, blue-leafed plants via tall, green-leafed plants (especially those in Texas, which are predominantly green), to smaller plants with greener, shorter, stiffer leaves and a tendency to have multiple branched stems. The latter plants are considered by many authorities to be a different species – *Yucca thompsoniana* – but the borders between the two are ill-defined. My friend Daniel from Texas has spent a lot of time studying these in the wild and sees them as separate species that hybridise when growing together, thereby showing intermediate forms. *Yucca thompsoniana*, he tells me, will apparently hybridise with any yucca species found across its range. Some, however, regard them simply as geographical variants of a widespread and variable species. In any event I include them both here as there is clearly an affinity. The European nursery trade gives a simplified, if skewed, perspective as plants are selected from the extremes of appearance. Tall, single-trunked (rarely branched), blue-leafed plants that have been neatly trimmed of their old leaves are sold as *Yucca rostrata*; whereas smaller, much branched plants with smaller, greener leaves and a full petticoat are sold as *Yucca thompsoniana*. The reality is not so clear-cut and, although I have no strong views either way, the smaller *Yucca thompsoniana* plants are more cold-hardy, taking down to –20°C.

Both *Yucca rostrata* and *Yucca thompsoniana* are fairly widely grown in the UK but they are a lot harder to please than *Yucca linearifolia*. When happy they are stunningly beautiful. Full sun and perfect drainage will help them towards that goal.

Tall, green form of *Yucca rostrata* in Texas.

Shorter, multi-headed plant of *Yucca thompsoniana* in Texas.

Stiff and compact crown of Yucca thompsoniana.

Many growers consider it better to start with seed-raised plants, rather than the larger trunked specimens, as they settle in more quickly, not suffering from any setbacks at all and, in the case of single-trunked plants, can even outgrow plants that were initially taller but with less-established roots. One thing I would add, however, is never give up on a plant that appears to have died until it actually disintegrates. I had a large, bald, rootless trunk given to me by someone who did not have the space to keep it on the off-chance of it recovering. I kept it in a pot filled with nothing more than some all-in ballast, tucked it down the embarrassingly untidy side of my greenhouse (doesn't everyone have one of those?) and promptly forgot all about it. Maybe three years later I went to move it and discovered the crown was starting to regrow, so I let it be. Two years later there were roots appearing from the drainage holes in the pot. It is now growing away happily in an open sunny spot.

An impressive *Yucca × schottii* at East Ruston Old Vicarage.

A neatly manicured *Yucca × schottii* at The World Garden.

A closer look at the crown of *Yucca × schottii.*

Yucca × schottii

Common name: mountain yucca, Schott's yucca
Distribution: Arizona, New Mexico, across the border into Sonora and Chihuahua
Hardiness: H6

Studies have shown *Yucca × schottii* to be a naturally occurring hybrid between *Yucca baccata* and *Yucca madrensis*, forming stable, reproducing populations that spread naturally in its habitat. It is found at high altitudes, well over 2,000m, and hence adapted to cooler growing conditions than most. It is a chunky thing, with a 30cm-wide trunk that can be anything up to 5m in height and occasionally branched. Broad, sword-shaped, pale grey-green, sharp-tipped leaves, around 60cm long are often held for a long time, clothing the stem a long way down. Offsets will often grow from the root stock, eventually leading to a large clump. White flowers are produced on a short panicle in summer.

It is to be found in wooded rocky canyons in the mountain ranges that extend around where the states of Arizona and New Mexico meet with the Mexican states of Sonora and Chihuahua.

In cultivation in the UK this is both cold-hardy and more moisture-tolerant than many yuccas and can actually grow quite rapidly. A friend planted a small, 30cm-high plant and eight years later it had grown to 1.8m high, fully clothed with leaves down to ground level, taking down to −10°C. I transplanted this very plant five years ago from its position in full sun into a semi-shaded spot where it languished markedly. A year or so ago I moved it again to a position in full sun and it immediately perked up and seems to have taken this latest move without issues. In nature it will see far lower temperatures, down to −20°C. The roots contain high levels of saponins, which are toxic to cats and dogs, so if yours is an inquisitive animal that scrabbles at plants it is safest to ignore it and plant something else.

Yucca torreyi

Common name: Torrey's yucca
Distribution: New Mexico and western Texas
Hardiness: H6

This is yet another species where the names are a complete muddle. Some authorities would have us believe these are synonymous with *Yucca faxoniana*. Others – the Kew checklist included – tell us we should call them *Yucca treculeana*. I grow plants of all three taxa – big plants, all of which have flowered – and I cannot for the life of me see how any of them can be considered to be the same, as the leaves, stems and, critically, flowers of each are significantly different. Something in the paper trail of published names might indicate synonymity but certainly not the actual plants. So I will stick with *Yucca torreyi* and suffer the consequences.

My *Yucca torreyi* has grey-green, deeply guttered leaves that grow to around 80cm long with conspicuous marginal filaments. The tips of the leaves end in very sharp spines: I perhaps thoughtlessly planted it near a main pathway and so have impaled wine bottle corks on to the spines to prevent injury or lawsuits. In 30-odd years from seed it has now grown a 1m-tall stem that is fairly slim but rigid, around 20cm in diameter. It has flowered on four occasions and in my view is one of the most handsome of all yuccas in flower. The bud arises from deep in the crown early in spring, resembling an aubergine, slowly expanding then gradually opening up to a many-branched stemless panicle held within the upright leaves. The flowers open from maroon buds to reveal gorgeous nodding, creamy white bells that still have maroon on the reverse side. It really is spectacular. After flowering, a new growth bud appears next to the old flower stalk, sometimes alarmingly late. It seems to then develop another full crown over the course of three or so years before flowering again.

In habitat the plants grow in open plains and rocky slopes and can often look quite scruffy, reaching 6m in height, sometimes branching at the stem, sometimes offsetting from the base, sometimes, both. They will take down to –15°C; certainly my plant has taken my worst winters without issue. If you can find a spot for this where the nasty leaf tips are out of harm's way, it is certainly worth growing for its distinctive, slightly dishevelled look and stupendous flowers.

Yucca torreyi flowering impressively.

Big, fat, colourful bud of *Yucca torreyi* that is full of promise.

Closer look at the beautiful flowers of *Yucca torreyi*.

Worth Considering

Yucca elata

This is a beautiful species from extremely arid parts of Arizona and Texas, with dense crowns of very narrow, dark grey-green leaves, with attractive white margins. Mature plants in habitat can reach 5m and have many-branched trunks. It is a difficult plant to grow well in the UK. Imported trunks have generally had the deep roots broken off, a trauma from which they rarely seem to recover. Seed-raised plants soon start to suffer by being confined in a pot. When everything works, however, this is a hard species to beat in looks. They are hardy to –20°C. (H6)

Yucca elata at East Ruston Old Vicarage.

Yucca elephantipes

More accurately *Yucca gigantea* but almost universally known by its old name, this is one of the more tropical species, probably most familiar as a houseplant, that grows a huge elephant-foot base to its trunk and has relatively soft leaves. It is surprisingly hardy, taking –6°C without much damage and recovering from –8°C in my garden. If covered or growing in a particularly good microclimate, they can take lower temperatures. There was an immense specimen growing in front of the south wall of a house along the A13 at Westcliff-on-Sea in Essex that I used to drive past regularly. It had been growing there for two decades to my knowledge, through all the worst winters this area has seen, and the tops of the leaves had reached beyond the height of the roof eaves. A few years ago the house changed hands: evidently the new owner did not share my admiration for the plant as it is no longer there. For the milder counties it is a quick-growing and spectacularly impressive plant. There are long-leaf and short-leaf forms; I am growing both at the moment and believe the longer leaf form to be a tiny bit hardier. It is also sometimes seen offered under another old name *Yucca guatemalensis*, which may apply to the short-leafed form only (or not). (H3/4)

Yucca elephantipes in the snow.

Yucca elephantipes in a front garden in Shoeburyness, Essex.

Yucca filamentosa in Richard Darlowe's garden in Barnsley.

Yucca filamentosa

This is one of the trunkless species that, in time, forms a cluster of fairly rigid rosettes with dark green leaves with marginal filaments. It is actually quite hard to find the real thing in the UK – mostly the trunkless yuccas seen for sale are scruffy, lax plants that are more akin to *Yucca flaccida*, another stemless species but of less elegant form. Both carry impressive flower spikes in summer, are not hugely fussy about soil and are extremely hardy. (H5/6)

Yucca glauca

Another trunkless species, this one grows into clusters of rosettes comprising scores of extremely narrow grey leaves with threaded margins. It is fussier than the previous two species, being far more xeric in its requirements. If you can keep it happy it is hardy and elegant, making compact tight domes of shimmering narrow leaves from which emerge tall and usually unbranched spikes of showy white flowers. (H6)

Yucca queretaroensis

This is arguably one of the most beautiful succulent plants it is possible to grow. It is clearly a close relative of *Yucca linearifolia*, growing a similarly impressive domed crown comprising hundreds of four-angled, stiffly held, exceptionally narrow,

Yucca queretaroensis.

green leaves. When lit from the side by afternoon or winter sun, the whole head seems to glow and shimmer: a perfect study in light and shadow. Trunks are similarly clothed in a tight-fitting petticoat of dead leaves. In nature they cling in clusters to the sides of almost sheer ravines, around the Rio Estorax in Querétaro in the middle of Mexico, tucking themselves out of harm's way. It is generally suggested that these are exceptionally rare but so far as I have seen, there are plenty of scattered populations. I have visited a couple of groups that are accessible via vertiginous roads cut into the sides of the ravines, allowing a close glimpse of these spectacular yuccas in their equally spectacular habitat. I have a wonderful specimen that has taken my worst three winters, although I have read that hardiness tails off much below –12°C. A proportion of plants raised from seed show hybrid characteristics, the main suspect for the other parent being *Yucca filifera* with which it grows in its habitat. These are proving to be good garden plants, fast-growing with a habit somewhere in between both parents. (H4)

Yucca queretaroensis catching the light in its sheer-sided canyon home.

Yucca rigida

This is one of the most enigmatic and frustratingly fickle plants I have ever grown. These are superbly architectural yuccas having a similar pale blue colour to *Yucca rostrata* but with a much thicker, wider and more rigid leaf. I have tried and failed to grow this many times; whether a trunked plant or seed-raised plant with a full rootball, none ever took. I had one last-ditch attempt – an established plant from a Sicilian nursery that had been raised from seed, badly neglected but fully rooted. I kept it in my unheated greenhouse, waiting for the right place to plant it, but all the while the plant was going backwards, the crown reducing in size. The opportunity finally came round in my new garden extension; I was able to plant it in full sun in nothing more than builder's rubble – concrete and house bricks mixed with crushed mortar and

Yucca rigida enjoying its change in fortune in my rubble garden.

A huge *Yucca rigida* in the Desert Botanic Garden, Arizona.

Yucca treculeana, short-leafed form.

Yucca treculeana, longer-leafed form, in habitat in Tamaulipas.

plaster from the demolition of an old house. In the course of two years it has completely turned around and is now growing happily and healthily, regaining something like a full crown again. At last! This is a plant worth persevering with as it has such a strong presence. It seems all you need to do is simply demolish a house and plant it in the wreckage. (H5/6)

Yucca treculeana

This is another impressively chunky species of yucca. There are short-leaf and long-leaf forms available – I grow both and they are equally spectacular, perhaps the longer leaf form the more dramatic. Plants grow a stout, branching trunk, usually dividing after flowering. Leaves are rigid (especially on the shorter leaf form) with a strong spine on the tip. There are no marginal leaf filaments. Flowers are impressive, an upright short-branched panicle, tinted maroon to start, quickly filling out with large white bells that sometimes retain a pale mauve reverse. I have seen this species growing in north-eastern Mexico, often in light shade. I tried planting one in a similarly shaded place in my garden but it has never grown very well, becoming lax and lacklustre, so I conclude that in the UK a position in full sun would suit it best. My biggest plant has taken all of my worst winters without issue and I imagine this species would be hardy to temperatures around –15°C. (H5)

Yucca baccata in habitat in Arizona.

Small plant of *Yucca baccata* at The World Garden, Lullingstone.

Watch List

Yucca baccata

This is a short yucca, growing a little trunk in time, that is extremely drought-, heat- and cold-tolerant but a little fussy when it comes to wet feet. Leaves are deeply guttered and strongly recurved at first, changing as the plant ages to become straighter and flatter, and have an exceptionally coarse texture. I am sure you could strike matches on them. I have never managed to keep one of these going for long, losing my biggest and oldest plant one wet winter. I aim to try again as it makes an impressive lump of a thing. (H5/6)

Yucca pallida

This is an absolutely gorgeous trunkless species with reasonably soft, wide, china-

Yucca pallida.

The large, yellow flowers of *Yucca pallida*.

blue leaves, growing slowly into an open rosette. Large pale yellow bell-shaped flowers are produced on a 2m-high panicle, after which the plant retreats underground (at least mine did) to regroup before growing back as a small cluster. A very charming small species that is rather different in appearance to most yuccas and worth seeking out. Hardy to –10°C or lower if covered. (H4/5)

Yucca periculosa

At first glance *Yucca periculosa* is very similar to *Yucca filifera* but is found growing further south, taking over the same sort of niche, in the states of Puebla and Oaxaca. It grows with age into a massively branched monster, perhaps not as scruffy as *Yucca filifera*, and with upright, not pendulous, flower spikes. Up close the newer leaves are a glossy mid-green, with curling marginal filaments and, when young, distinctive red margins that fade with age. I have a plant in my garden that I grew from seed eighteen years ago that is now 1.5m high with two trunks. It has taken all the bad winters during this time and been transplanted twice. It is now in its final resting place, so hopefully the growth will settle down and speed up. I find it very attractive for its glossy green leaves – a welcome change to the usual glaucous forms. (H4)

Yucca periculosa in its habitat near Puebla.

Yucca periculosa, crown detail of my plant.

Yucca potosina

This species is another Mexican yucca that is not very widely grown. Young plants are almost impossible to distinguish from *Yucca filifera* but, at maturity, they are only sparsely branched and, in habitat at least, a little more easily distinguished. I have had one growing in the garden for around twelve years. It started off in a spot where surrounding plants quickly swamped it to the extent that I had actually forgotten it was there. I stumbled upon it again two years ago when cutting the other plants back, rescued it and settled it into an open, sunny position. I do not think it enjoyed the move and during my early cold winter the stem rotted through and fell over with no sign of recovery. (H3/4?)

CHAPTER 7

OTHER SUCCULENTS

SUCCULENT ASPHODELS

Aloe

Family: Asphodelaceae
Subfamily: Asphodeloideae

Aloe is a massive genus with over 600 species of succulent plants distributed throughout Africa, Madagascar and the Middle East. They range from massive trees to tiny ground-hugging rosettes and many species, notably *Aloe vera*, have been used throughout human history in folk medicine for applications as diverse as soothing burns and relieving constipation. Taxonomists have recently been at work within the genus, relegating many of the more familiar plants to different genera, most of which I have reluctantly adopted but kept within the broad heading of *Aloe*. Considering it is such a huge, widely distributed group, it is perhaps disappointing that there is only a scant handful of reliably hardy species. There are, however, several more species flirting around the edges of hardiness that are worthy of discussion.

Keystone Species

Aloe polyphylla

Common name: spiral aloe, *Kroonaalwyn*
Distribution: Lesotho, South Africa
Hardiness: H6

Aloe polyphylla is considered by many to be the Holy Grail of hardy garden succulent plants. It is a rosette-forming leaf succulent that grows into the most surreal of forms, a perfect spiral, reaching up to 70cm in diameter. The grey-green leaves are short and triangular with stubby teeth along the margins and, on the underside, an offset keel that effectively makes the leaf lie at a funny angle. When the next leaf grows it is forced to lie slightly to one side, which, upon repetition, forms the foundation of the spiral growth. Spirals can be clockwise or anti-clockwise, and rosettes are almost always solitary. The leaf tip can die off and wither, turning black and adding to the fascinating form and structure of this enigmatic beauty. After a few years, one or more branched flower stems are produced in summer, generously supplied with salmon-red tubular flowers.

In nature, plants are restricted to just one or two locations at the very top of the Drakensberg Mountains in Lesotho, a tiny annexed country within South Africa, at altitudes of up to 2,600m. Until recently it was considered to be extinct in the wild but a few have been rediscovered. They grow in coarse scree pockets on rocky slopes. In winter, plants are covered in snow and endure freezing temperatures down to −15°C. Summers are cool, misty and cloudy, rainfall levels are high at over 1,000mm annually, plus the freely draining substrate is constantly washed by cold snowmelt. It would be fair to say the growing conditions are challenging, more akin to those of an alpine plant than a succulent.

Luckily plants respond to pot culture quite well and there are enough large, regularly flowering specimens in circulation to ensure a ready supply of cultivated seed. This is a good thing because I just keep killing them in my attempts to grow one successfully – I have lost track of exactly how many. Nevertheless other people, all over the length and breadth of the country, have managed quite happily. It seems the most

OPPOSITE PAGE: Orange form of *Aloiampelos striatula* on St Michael's Mount.

Aloe polyphylla on St Michael's Mount.

Aloe polyphylla on a hot dry slope at Ventnor Botanic Garden.

Perfect spiral of *Aloe polyphylla*.

Snow-covered *Aloe polyphylla*.

successfully grown plants are potted or, if planted out, given a cover for winter. I know of plants outside that have taken temperatures down to –10°C, been covered in snow and ice yet still carry on happily. Each grower seems to have their own tip for success. Some swear by bark chips as the growing medium, another will say crushed brick. Some swear by watering their plants each time they walk by, others just leave them be completely. I just swear when I see yet another attempt to grow this perplexing plant has failed. I wonder if my area is just too hot or too dry? Anyway, I will keep trying.

Aloiampelos striatula

Common name: none applied
Distribution: Karoo, Eastern Cape Province, Lesotho
Hardiness: H6

Aloiampelos striatula, more widely known under its former name *Aloe striatula*, is a rambling, untidy shrub comprising many lax branches to roughly 1.5m high and broad, although older, unpruned plants can be more than 2.5m across. Leaves are around 30cm long, slightly recurved, guttered with weak, pale teeth along the margins. Where they clasp the stems the leaf sheaths are conspicuously marked with longitudinal lines, hence the name.

Aloiampelos striatula at RHS Hyde Hall.

Inflorescences are spectacularly pretty, being 40cm racemes of downward-facing, tubular, yellow flowers, orange in bud, with noticeably protruding anthers. A large shrub in full flower is quite a sight and positively drips with nectar.

There is one recognised variety, *Aloiampelos striatula* var. *caesia* that has greyer, shorter, more upright leaves and shorter flowers that are almost entirely yellow and held more horizontally on the spike. The stems of this variety are much more inclined to branch. I have a particularly striking form in which the flowers are more strongly coloured with orange and an, as yet, untested one purportedly with more bicoloured flowers of orange and cream. Old literature from the 1950s describes a red-flowered form from certain locations but to my knowledge this is not in cultivation, even in South Africa. If anyone reading this has one, please get in touch.

This is a plant from high mountains and is arguably the most cold-hardy of all the *Aloe* family. Top growth can get damaged at just –6°C but plants will almost certainly sprout from older, woodier stems. Temperatures around –10°C can cut it back to the ground but, again, plants will generally regrow from the woody rootstock underground. As outlined back in Chapter 1, my friend Mark Fillan's original plant survived a low temperature of –18°C, coming back from underground, during one of those punishingly cold winters of the mid-1980s. I have found it best to grow this as hard as possible in the rockiest, poorest soil available in full sun, which keeps the plant much more compact. For me *Aloiampelos striatula* var. *caesia* is a more attractive garden plant. The branching habit gives a denser and compact shrub, the flower colour is more pronounced and, comparing side by side with the regular form after my early and late cold winters, I would say, tentatively, that the top growth takes more frost before becoming damaged.

This could easily qualify as the perfect garden succulent plant but for a few issues. Both forms can be blighted by blackfly and their attendant ants, which, if not dealt with swiftly, will completely disfigure an entire season's worth of flowers in no time. Also there is a tiny snail, apparently active at night, that will happily rasp huge patches from the leaf surfaces leaving unsightly scars. Finally, plants are martyrs to their own vigour, soon outgrowing their allotted space and becoming untidy. The latter is easily fixed with secateurs and an empty green waste bin, although you will lose a year's flowers by cutting back hard. Nevertheless, it is one of the essential stalwarts of the UK xeric garden.

Regular flower colour of *Aloiampelos striatula*.

Yellow flower and greyer foliage of *Aloiampelos striatula* var. *caesia*.

Orange form of *Aloiampelos striatula*.

Aristaloe aristata

Common name: *Serelei*
Distribution: The Karoo, Eastern Cape Province, Free State, Lesotho, KwaZulu-Natal
Hardiness: H5

Aristaloe aristata is another victim of taxonomists' scrutiny, previously widely grown as *Aloe aristata* but now considered different enough from the rest to be isolated in its own lonely monotypic genus. With such a wide distribution, individuals will obviously vary quite a lot but, typically, it grows into 15cm-diameter rosettes, densely packed with triangular leaves that are covered on both surfaces with pale, toothy warts. Leaf margins are serrated with small, pale teeth and the leaf tip is often extended into a long, bleached and soft tip. Plants offset profusely, growing into football-sized colonies. When drought-stressed the whole plant curls into itself almost like a disturbed hedgehog. Flowers are abundantly produced in early summer, a branched raceme carrying up to 30 nodding, curved, coral-coloured, tubular flowers with a conspicuous bump at the base.

This species enjoys a huge natural distribution from lowland tropics to high mountain peaks with an associated wide range of environments. This accounts for the plant's adaptability to cultivation and popularity. I would hazard a guess that every single one of my gardening acquaintances has grown this at some time or another.

I have grown this in my garden for 25 years. My worst winter nearly saw it off and my early cold winter has caused it some damage, but it is still there. With a cover to keep the winter weather off, this will take down to –12°C.

Several named forms have been selected, the cultivar 'Cathedral Peak' being a personal favourite with its extra wide, dark green leaves and pronounced warty leaf surfaces. Around eight years ago I was given a particularly old clone,

Aristaloe aristata in flower at East Ruston Old Vicarage.

Nice chunky form of *Aristaloe aristata*.

grown from seed some 50 years earlier, that is most attractive with extremely short, wide leaves. I am also hugely excited to have recently been given a small plant from a known high-altitude provenance that has a slightly different look about it – the leaves are smaller, a nice blue-grey in colour. It seems to be a slow grower but eventually I should be able to test this outside.

There is a plant that is commonly seen for sale that goes by the name *Aloe aristata* 'Cosmo'. It is similar to *Aristaloe aristata* 'Cathedral Peak' but with broader leaves and fewer warts. It is more accurately believed to be a bigeneric hybrid × *Gasteraloe* in origin, although how the name change to *Aristaloe aristata* changes this hybrid name I cannot begin to imagine. Surely not '× *Gasteraristaloe*'? In any case, 'Cosmo' seems to be slightly less hardy than the *Aristaloe aristata* parent but will certainly take down to –5°C unprotected.

Aristaloe aristata 'Cathedral Peak'.

Aristaloe 'Cosmo'.

Aristaloe 'Cosmo' flowering on St Michael's Mount.

Worth Considering

Aloe arborescens

Only hardy to perhaps –3°C or a bit lower in the open garden but, if your climate allows, this is a lovely species to grow. It forms a many-headed shrub, something like *Aloiampelos striatula* but chunkier, neater and toothier, anything up to 1.5m high and across and, in late winter, pushes up poker spikes of bright red, tubular flowers. With a cover it can take lower temperatures but the flowers are damaged regardless of being wet or dry. In a mild winter they certainly add a welcome splash of colour. *Aloe mutabilis* is very similar in appearance, also flowering in winter, but the flowers are bi-coloured red and cream, and it is perhaps a degree or two hardier, albeit harder to come by. (H3)

Aloe arborescens on St Michael's Mount.

Aloe mutabilis at the base of my house wall.

Aloe mutabilis regrowth following defoliation at –5°C.

Aloe broomii

This forms a solitary agave-like rosette, to around 80cm diameter, of viciously spiny, olive-green leaves that curve inwards. In time it grows a short trunk and has a curious 'rat-tail' type of inflorescence of overlapping scales through which just the anthers of the flowers protrude. In theory, this should be pretty hardy as it comes from high mountains in the middle of South Africa. In practice, I have tried this many times and only succeeded in overwintering it intact on a couple of occasions, certainly never keeping it going for very long. I suspect it needs to be kept dry in winter and, to that end, my latest attempt is sitting right at the base of a *Phoenix canariensis* palm that gives it some measure of cover. The palm itself is planted under the canopy of a tall cypress

Aloe broomii at Tremenheere Sculpture Gardens.

tree so the whole area remains quite dry. So far so good, my early cold winter has been and gone and the *Aloe broomii* looks to be in good condition. (H4/5?)

Aloe ecklonis

This is one of a group of aloes known as 'grass aloes' that behave almost like bulbs, adapted to areas of frequent forest fires by regrowing from underground if the top is burnt off. This can, theoretically, work to our advantage in that if top growth is damaged by frost, it will still regrow. I grew this species for several years, finally losing it during my worst winter. It is not the sort of plant that appeals to 'regular' succulent growers so I have not encountered any replacements as yet and may even resort to growing some from seed. It makes small rosettes of broad, flat, green leaves and almost allium-like compact heads of salmon flowers.

Aloe boylei is similar. (H4/5)

Aloe cooperi is another grass aloe that is showier, larger but less hardy. (H3/4)

Aloe ecklonis.

Aloe maculata on St Michael's Mount.

Aloe maculata
(syn. *Aloe saponaria*)

This is part of a group of hard-to-distinguish speckled aloes that form flat, open rosettes of toothy-margined leaves, the surfaces of which are marked with dashes, dots, spots, speckles and stripes. They offset copiously, forming spreading colonies if the climate allows. Flowers are borne on a sparsely branched, flat-topped raceme and can be anything from yellow to, more commonly, salmon-red. If dry this will take down to –5°C without much damage and, if killed to ground level by colder temperatures, can sometimes come back from underground. I have never kept it alive for more than a couple of winters but put that down to never planting it in exactly the right place. Next time! (H3)

Aloe greatheadii and *Aloe fosteri* are superficially similar but have different inflorescence structures and, ultimately, are less hardy. Aloes are, however, a louche and promiscuous bunch: their world abounds with hybrids and these little speckly ones are among the worst offenders.

Aloe × nobilis

This is a fairly common plant in the houseplant trade almost universally wrongly called *Aloe mitriformis*, which does exist and is something similar but different. There has been a recent push from some taxonomists to lump these two, and all the other creeping aloes, together under the umbrella name *Aloe perfoliata*, which, if nothing else, saves all

Aloe × *nobilis* on St Michael's Mount.

Aloe brevifolia with leaf-tip damage after –5°C.

the angst over getting the right name. *Aloe* × *nobilis* grows into a cluster of chunky rosettes of mid-green, warty leaves with pale spines on the leaf margins, which, as they age, elongate and fall over to then creep along the ground. Flowers are borne on flat-headed panicles and generally red. If dry they can take down to –5°C or a bit lower but in the open garden mine have always been killed at that temperature. (H3/4)

Aloe brevifolia is ostensibly 'similar' at first; individual rosettes are of tightly packed, chunky but short, pale china-blue leaves with pale marginal teeth and a line of little teeth along the back surface. It does not fall over and creep as it ages, however, rather offsetting its way to becoming a neat tight dome. Flowers are borne on a pointed panicle. Beautiful as it is, a temperature of –5°C will kill it unless it is covered. (H3/4)

Kumara plicatilis
(syn. *Aloe plicatilis*)

Another victim of taxonomy, this is a stunningly beautiful plant that eventually grows into a medium to large shrub, in its habitat reaching 5m tall. The plants have fat, flattened, linear leaves in blue-grey that are produced in two opposite ranks, giving a two-dimensional fan shape. In the usual course of events, the plant will carry on growing for a few years, gradually forming a trunk. A raceme of coral-red flowers is produced from the centre of the fan, after which the plant

Kumara plicatilis in my greenhouse.

divides into two growth points and grows two fan-shaped heads. This dichotomous growth pattern continues: after each flowering, that head divides and the trunks continue to extend. The stunningly unique growth pattern aside, another interesting aspect of this plant is that it comes from a region that will see up to 1,200mm of rainfall in a year and it is also hardy to a certain amount of cold. My plant has taken –6°C with a cover, with minimal damage and I suspect it would go a bit lower than that. For milder, wetter counties this is definitely worth trying as it is a show-stopper at all ages. (H3/4?)

Watch List

Aloe humilis

This grows in a similar fashion to the more familiar *Aristaloe aristata* but the leaves are narrower, blue-grey but still with pale warts and teeth. Rosettes are incurved and, overall, plants are smaller, offsetting to form a tightly packed cluster of a dozen or so heads. Flowers are an unbranched spike of coral tubes. This does have some hardiness but needs to be kept dry – either covered or tucked away in a crevice with an overhang. I have never kept it going for more than four or five years but replacement plants are cheap and readily available.

Aloe marlothii

This is a hefty brute of a plant, a giant fleshy grey rosette of fat, incurving leaves covered on both faces and the edges with nasty spines; it eventually grows a trunk. If you can manage to keep this completely dry in winter it does have some hardiness to, say, –5°C, but it is a monster to work around. A friend who lives near St. Albans kept a large, potted specimen alive for many years by dragging it into the corner between two house walls and covering it with fleece, until one particularly cold year saw it off. *Aloe spectabilis* is similar and said to have a little more hardiness to cold. *Aloe ferox* is also similar, although it is less hardy.

Aloe marlothii in a Canary Island garden.

Aloe pratensis

This is a frustrating plant. In nature its range extends way up high in the Drakensberg Mountains and it is said to tolerate down to –10°C or lower if kept completely dry. It is a lovely thing, making chunky, blue-grey rosettes to 20cm across, offsetting to make a small cluster. Leaves are fat, triangular, with pronounced sharp teeth and warts. The flowers are huge for the size of the plant – the inflorescence stalk covered in papery bracts and the pale coral tubular flowers opening in rows from a fat, conical bud. I have tried it a couple of times in the past but never managed to keep it for more than three or four years. Recently it seems to have vanished from the horticultural trade: I have not seen plants available commercially or in collections for many years, although I have just managed to order some seed.

Aloe humilis.

Aloe pratensis.

Aloe reitzii in my greenhouse.

Aloe reitzii

As a young plant this strongly resembles young plants of *Aloe marlothii* but they soon change to being more slender. Narrow leaves, armed with toothy margins, curve inwards making neat, trunkless rosettes to around 80cm across. Flowers are an impressive branched candelabrum of tight red pokers. According to the Silverhill Seeds people in South Africa, this is a USDA Z7 plant, which would equate to roughly H5, so I keep persevering with it, trying it in different situations. No luck so far, my trial plants barely scrape through mild winters. One day I will hit on the perfect spot!

Aloe speciosa

This is an impressive plant, growing a fairly unruly-looking rosette of thick but relatively slender, grey-green leaves to 90cm long and quickly forming a trunk that in nature reaches a whopping 6m high. A charmingly captivating quirk of the plant is that the rosette tilts towards the light. The inflorescence is produced in late winter, an impressively fat rat-tail of pinkish-red buds that open to green, tubular flowers sequentially from the bottom of the spike, starting on the side facing the light. Historically I knew of a plant of this species growing in south-west Cornwall, but that is a different climate to here in my south-east garden, so I largely discounted it. Recently, however, I was given a magnificent old plant, over twenty years from seed, that had spent the previous few winters outside in its pot, dragged close to the house wall. Encouraged by this I planted it in as good a spot as I could muster, under the cover of a high evergreen canopy with its feet in rubble, and it has done spectacularly well. My early cold winter saw off a lot of experimental (and established!) aloes but this came through showing only leaf-tip damage with just a single layer fleece bonnet by way of protection. The emerging flower bud was undamaged. This is an immensely exciting prospect; a huge garden aloe that takes cold has to be one of the most desirable xeric garden plants.

Aloe speciosa flower detail.

Aloe speciosa.

Aloe × salm-dyckiana

A naturally occurring hybrid between *Aloe arborescens* and *Aloe ferox*, this actually looks just as you would imagine it, halfway between the two parents, growing into a branching shrub with chunky fat, grey leaves that have marginal teeth. Flowers are held on branching candela-

Aloe salm-dyckiana.

bra and are erect, narrow spikes of deep red. Surprisingly I have found it hardier than either parent and, kept dry, I was able to grow it for a number of years at the base of the south wall of my house, where it withstood temperatures down to –5°C. Branches and basal offsets are easily removed and rooted for insurance as it is never going to be a long-term prospect in anything but the mildest garden, but fun while it lasts.

Aloiampelos ciliaris

This is a lax scrambling plant, in nature twining its way through shrubs, with leaves that clasp around the slender, striped stems with almost hairy, fringed margins. It is related to *Aloiampelos striatula* but slimmer and more untidy. Inflorescences are relatively short but have colourful spikes of bright orange-red, tubular flowers that have yellow tips. It is said to be hardy down to –7°C but I have never been able to keep it alive when temperatures dip below around –4°C or so. I will keep trying as the flowers are absolutely spectacular; in fact, I have a space in mind right this minute.

Aloiampelos ciliaris in Ventnor Botanic Garden.

Aloidendron 'Hercules'

This is a man-made hybrid between *Aloidendron barberae* and *Aloidendron dichotoma* that is a monster of a plant, growing into a small tree. Everything

The astonishing *Aloidendron* 'Hercules' in Darren Hobbs' garden.

about it is attractive, from the deep green, upright, guttered leaves to the patterned trunk: smooth when young, attractively fissured when older. It is said to take down to −7°C in cultivation in the USA. In this country, plants for sale are few and far between and extremely expensive when they appear and, therefore, not conducive to risky hardiness trials. My friend Darren Hobbs, however, has been growing one of these plants in his garden in Newport, South Wales, for over a decade. He goes to great lengths to keep it dry, by covering it with a temporary greenhouse, and frost-free by insulating it with old sleeping bags, never leaving anything to chance and something of a Herculean task. By doing so he has kept it safe and snug when outside temperatures dropped down to −9°C. The plant now stands at over 3m tall, a testament to his determination and willingness to think outside the box.

Kumara haemanthifolia

I list this mainly in the slim hope that I might one day find another, having killed the single precious specimen I was given twenty years ago. It is related to the shrubby *Kumara plicatilis* but is stemless, growing its colonies of fan-shaped rosettes at ground level, strongly resembling a patch of *Haemanthus* bulbs. In nature, it grows at extremely high altitudes in high winter rainfall areas, favouring the colder, shady side of the rocks, and surviving freezing temperatures and occasional fires by dint of an extremely large underground rootstock. All of which sounds incredibly encouraging; however, it is reputed to be very difficult to grow in cultivation. Mine certainly was. So very few people have tried to grow it in the UK that it is difficult to say what it needs for successful cultivation. I have only ever heard of two people who have ever owned it – me and the person who gave it to me – but I am eternally optimistic.

Bulbine

A South African genus of 80-odd species, one of which is worth trying in the garden.

Hybridising for Cold Tolerance

As a general rule I prefer my plants to be as nature intended: either species or natural hybrids. My principles, however, can be easily swept aside when it comes to aloes. So few *Aloe* species are reliably hardy but there are many that are tantalisingly close to being so: the possibility of creating hybrids with the specific aim of good cold-tolerance is so very tempting. Most aloes will flower each year once they reach a certain size, so there is potential to get quite a regular breeding programme going. Crossing the hardiest species, such as *Aloe polyphylla*, *Aloiampelos striatula* and *Aristaloe aristata*, with, first, each other and, second, borderline hardy plants, could give extremely interesting results. Some have been made already. Many years ago, a friend of mine crossed *Aloiampelos striatula* with *Aloe ecklonis* and one of the seedlings grew into a very handsome plant that took a number of winters before succumbing to around −10°C in the winter of 2009/10. More recently in South Africa someone has crossed *Aloe polyphylla* with *Aristaloe aristata* and the result is a completely spectacular plant, spoken of in excited whispers in dark corners. With an increasing number of *Aloe polyphylla* now reaching flowering size, using this as a parent is becoming progressively viable and hopefully someone in the UK will successfully reproduce this cross. Any of the 'Watch List' species are worth trying, in my opinion: the main stumbling block as I see it would be having a pool of potential parents of flowering size that are kept frost-free, allowing flowering to take place unhindered by cold. Of course things do not always work out as expected. Many years ago I was given a plant of *Aristaloe aristata* × *Aloiampelos striatula* – potentially one of the very best – that was both ugly and not remotely cold-tolerant. Which is not to say that, if attempted again, the cross would end up so disappointing. Such is the nature of hybrids.

Aristaloe aristata × *Aloiampelos striatula* that only a mother could love.

Bulbine frutescens

is a small, branching, shrubby plant to around 40cm tall with thin, wiry stems and fleshy, almost chive-like leaves bearing, in spring, tall spikes of starry, yellow flowers. There is a cultivar, 'Hallmark' that has two-tone orange and yellow flowers, which is more common in the nursery trade than the plain yellow. Both can take some frost if dry. I have kept this going for several years in my garden wedged in between rocks so that the roots and stems are out of harm's way, taking down to around −5°C in this fashion, but finally seen off during my early cold winter. With overhead cover I believe they would take slightly lower temperatures than this. (H4)

Bulbine frutescens 'Hallmark' on St Michael's Mount.

Gasteria

A genus of around 30 (or 100, depending upon the authority) species of leaf succulents from southern Africa with hard but brittle, dark green leaves, often beautifully marked, named after the stomach-shaped swelling on the flowers. Amazingly, one of the species, *Gasteria excelsa*, has a decent track record for hardiness outside. It is one of the largest gasteria, with fat, sparse rosettes of chunky, smooth, triangular, dark-green patterned leaves with a serrated margin and a sharp keel on the underside, reaching 60cm or so across and offsetting to make a small cluster of plants. In summer, an outrageously long and spindly branched inflorescence is produced – mine has reached 1.8m tall – bearing those odd pink and green tubby flowers that dangle from the flowering stems like a string of lanterns. In habitat this species grows on steep, shaded cliff faces or in dense scrub, often at fairly high altitudes, in areas with moderate rainfall and where it will routinely experience frost. I grew a plant at the base of a yew tree for many years, losing it during my worst winter. A replacement has now been growing tucked away at the very base of the south wall of my house for eight years, taking my early and late cold winters without any issues. If completely dry I am sure this would take down to –8°C or lower. It is so unusual that it is definitely worth trying to find and grow, although large plants are generally quite expensive. *Gasteria* will tolerate a little shade so it can easily be grown at the foot of a larger plant that will give some overhead cover. (H4)

I have recently read that the similar but even larger species, *Gasteria acinacifolia*, has been known to survive down to –7°C in cultivation in the USA, despite being a coastal plant and, having managed to source a plant, I will use it for future trials. (H3/4?)

Gasteria excelsa.

STONECROPS

Aeonium

Family: Crassulaceae
Subfamily: Sempervivoideae

This is a genus of around 30 species native mainly to the Canary Islands but also other Macaronesian islands; one or two even stray on to mainland Africa. They are rosette-forming plants that, varyingly, branch into shrubs, remain on a single stem or remain stemless at ground level. They are generally winter growers, remaining dormant in the hotter, drier months: new growth comes with cooling temperatures and wetter weather, leaving them particularly susceptible to early frosts. None are exceptionally hardy but a few are worth trying outside, especially in the milder, wetter parts of the UK.

A good way to grow these is to have a cache of 'mother' plants, tucked safely away under glass somewhere, from which individual rosettes are cut. These cuttings can be poked directly into rock crevices or palm trunk leaf pockets where they will soon root and quickly grow. In a mild winter they will carry on into the following year but, even if frost kills them off, there will still be material available from the mother plant. Most of the species, hybrids and cultivars can be grown outside in this way, not just those with some hardiness.

Worth Considering

Aeonium aureum
(syn. *Greenovia aurea*)

This is an interesting plant, found at higher altitudes in the pine forests of the Canary Islands where it frequently hybridises with other nearby *Aeonium* species. This is a stemless species, forming a neat rosette of relatively thin leaves. It develops an interesting tightly packed, summer-dormant bud, all outer leaves drying off. When ready, the centre pushes out to a flat head of yellow flowers, after which the plant will usually offset, gradually forming a cluster. This has survived

Aeonium aureum.

–5°C in my garden – the only aeonium I grow that survived my early cold winter. (H4)

Aeonium simsii

This species comes from Gran Canaria, growing at the very highest places, such as Roque Nublo, the cloudy crag, or Pico de las Nieves, the snowy peak. These evocative names give clues as to the weather and give a definite impression that this species can tolerate cold: indeed, one time I visited Roque Nublo in November but had to abandon the walk due to horizontal sleet, freezing fog, and snow and ice lying on the ground. It grows in much the same way as a house-

Aeonium simsii.

leek, forming a mound of densely clustered heads each comprising pointed leaves that have cute marginal filaments and little striations on the reverse – attributes it passes on to hybrids, as well as cold tolerance. I have read that these do grow in sunny places but as yet all the plants I have seen grow in shade on the north side of rocks or at the base of tree trunks. In summer the whole plant condenses down to a dormant bud. I have grown this in my garden, on and off, for many years but lost it completely to my early cold winter, the rapid change from active growth to –6°C being too much. (H3/4)

Aeonium spathulatum

This is found on all of the higher mountainous Canary Islands, usually up in the pine forest but also out on open lava fields. It is an attractive plant that forms a small, much-branched shrub with small rosettes at the tip of each branch, rarely more than 2cm across. These rosettes close up to tight buds for their summer rest but, just before this happens, older plants may flower, producing a much-branched orange stalk from which the yellow flowers emerge in a kind of haze. I grew this in my 'cliff face' of lava stone and it must have felt at home as, for two consecutive years, it self-seeded in among the cracks. My early cold winter, however, killed it stone dead. (H3/4)

Aeonium spathulatum.

Aeonium × barbatum

A hybrid between the two previous species, in many ways this combines the best of each parent. It is free-flowering, making a small shrub with larger individual rosettes than *Aeonium spathulatum* and inheriting the cute leaf filaments and

Aeonium × *barbatum*.

striations from *Aeonium simsii*. It also seems a bit tougher than either parent, having endured my worst winter and late cold winter just fine. After 20 years, however, my early cold winter killed it. (H3/4)

Aeonium 'Maximus'

A fairly recently raised hybrid between *Aeonium cuneatum* and *Aeonium simsii*, this has vigour and a good deal of hardiness, taking my late cold winter in its stride. I grew it tucked into the pockets left by the stubs of old leaves on my *Phoenix canariensis* palm and it seemed very happy growing that way, right up until my early cold winter killed it. (H3/4)

Aeonium 'Maximus' growing on a palm trunk.

Adromischus

A genus of 25 or so leaf succulents from South Africa, a couple of which have shown some hardiness. *Adromischus cooperi* grows at high altitudes and has taken –5°C in my garden when dry; best grown in a rock crevice with some cover if possible. It makes a really interesting huddle of beautifully mottled, fat leaves, the leading margin of which is wavy, like the crimped edge of a Cornish pasty. *Adromischus trigynus* has flatter, oval, disc-shaped leaves that are, again, beautifully marked, taking on maroon tints in the stress of summer drought. With me, neither have been long-lived plants, a harsh winter has seen them both off, but they are easily propagated from leaf cuttings. (H3/4)

Adromischus cooperi.

Adromischus trigynus.

Cotyledon

A genus of some 35 species from southern Africa and Arabia, growing generally as shrubby succulent plants of which one, *Cotyledon orbiculata*, has an amount of hardiness. This has a wide range and is variable in appearance: typically a low, branched shrub with fat, oval leaves. The best forms have a thick silvery patina and a thin red margin to the leaves. Flowers are a showy nodding cluster of orange bells held on a central stem. I have grown this a few times and have finally reached the conclusion that it can take a reasonable amount of cold down to –5°C or lower, but only if covered. Frost or snow falling on to the plant will kill it. Another issue is that snails seem to find it irresistible and can spoil the look of a plant overnight. A sensible way to grow it would be to keep a mother plant, as with aeoniums, somewhere frost-free from which cuttings can be taken and grown outside in a variety of positions to find the optimum place. I may even take my own advice on this one year. (H3/4)

Cotyledon orbiculata.

Crassula

Crassula is a large genus of around 200 species of succulent plants from South Africa of which only one is reliably hardy, with a further two showing an amount of cold tolerance. *Crassula sarcocaulis* is an interesting plant, growing into a 60cm-high, many-branched shrub with fleshy stems and simple peg-like leaves that always make me think of a bonsai-trained hebe. Small, scented, pink flowers (there is a rarer white form, too) open at the tips of the branches in summer. This is an extremely hardy plant, taking down to –12°C without protection. The biggest problem with growing it is siting it so that the brittle branches are kept out of harm's way from accidental damage. My other issue with this is that it is rather scruffy and, if I am honest, a little bit dull. (H4/5)

Crassula sarcocaulis on St Michael's Mount.

Crassula multicava.

Crassula pubescens subsp. *radicans.*

Crassula multicava is rather prettier, in my opinion, and has some hardiness. Frost will kill the top growth but my plants have grown back from underground after taking –4°C. Stems are fairly brittle and bits fall off all the time, giving a ready source of cuttings. The inflorescence is borne on the stem tip, a wafty cloud of small, white flowers emerging from pink buds, after which the whole plant is weighed down with scores of bulbils, which, again, provide a source of back-up plants. (H3/4)

Rather different in looks, *Crassula pubescens* subsp. *radicans* is a low-growing, spreading plant that grows into a mat of 1cm-long, spoon-shaped leaves. These are apple green during winter, flushing coral-red when stressed through drought in summer at which point little bobbles of white flowers are also produced. I grew this for a few years, the plant taking down to –4°C, but lost it after an extended drought. I aim to try it again. (H3/4?)

Dudleya

These are rosette-forming succulents from the south-western USA, many of which are worth trying in the UK. The main issue is that they are summer-dormant, often reducing down to a small, resting bud and resuming growth again in cooler, wetter weather. As with aeoniums, this leaves them susceptible to damage by early frosts, which does seem to limit the choice. Of the species I have tried, there is one that stands out: I grew *Dudleya pulverulenta* in my vertical rock wall for several years, during which time it never failed to put a smile on my face. It formed a flat rosette of short, wide leaves to a massive 30cm across, the grey leaves generously covered in mealy white powder to such a great extent that excess powder accumulated at the base of the leaves, resembling a colony of mealy bugs. Rain would wash some of the powder on to the rock below, leaving a chalky residue. If that was not enough, a 50cm-long, branched flower stalk of silver buds, opening to cherry-red flowers, just added to the drama. My plant took several winters, including my late cold winter, but died for no apparent reason one summer, never coming out of dormancy but fading away. (H3/4)

Dudleya cymosa is smaller and less impressive but substantially hardier and easier to grow. Small rosettes of pointed, powder-coated leaves produce branched pink stalks carrying masses of showy, starry, orange flowers. It disappears to almost nothing in summer but has returned reliably each year for eight years or so. *Dudleya lanceolata* is rather similar with even more slender, less glaucous leaves; flowers on mine are yellow but this can vary. *Dudleya caespitosa* is, again, similar but fatter with denser rosettes, growing a short stem. Flowering stalks are dark pinky-red and flowers are pale yellow. I have not grown this for as long as the other species mentioned, maybe four years, but it is trying hard to impress. These last three should be hardy to –8°C or lower, best grown in perfect drainage in a rock crevice. (H4)

I grew *Dudleya edulis* for a few years; it has a different look to it, the leaves forming something like a cluster of slender fingers, or even an open sea anemone,

Dudleya pulverulenta.

Dudleya caespitosa growing on a palm trunk.

reaching out of the rocks. It survived my late cold winter just fine but was strongly summer-dormant, almost completely disappearing, and simply did not reappear after an extended drought. I would like to try it again.

One of the absolute gems of the genus is *Dudleya brittonii* but, after numerous attempts, I have given up as it is either more tender or just plain fussy.

Echeveria

Family: Crassulaceae
Subfamily: Sempervivoideae

Echeveria is a large genus of 150 or so species, mainly from Mexico but spreading north and south from there, with hundreds of hybrids and cultivars. They are popular as potted plants, although vine weevil can be a nuisance, and are widely grown. Some of the species show good cold-tolerance. These are often found in nature growing in cloud forest environments and in many ways are more suited to the milder, wetter south-west of the UK than the dry south-east, so I am sure there would be many more worth growing. I list here some of the ones I have grown.

Keystone Species

Echeveria elegans

Common name: Mexican snowball
Distribution: central Mexico
Hardiness: H4/5

This is a very pretty species from mountainous parts of central Mexico. Rosettes are around 15cm across, comprising many tiers of the loveliest china-blue leaves that curve inwards along their length and breadth, leading to a charming wide goblet effect. Offsets are freely produced and, in no time at all, a single plant soon becomes an undulating, crowded colony. In late winter the leaves take on a pink tinge, making the whole clump attractively colourful. Nodding pink flower stems are produced in early summer, buds opening to show their yellow flowers.

Echeveria elegans showing winter tints.

Echeveria elegans as ground cover.

I have seen this growing in several locations in Mexico: always around 3,000m altitude, way up high, where mists, fog, rain and snow are common. It is most certainly not a desert plant. I have seen it growing equally happily in pockets between rocks, over the top of rocks, on the trunks of trees, in among grasses; it is a very adaptable plant in nature.

I bought my first plant around 30 years ago and ever since then I have grown it outside, unprotected, and have never lost it; the plants took my worst three winters with little or no damage. I have been spreading it around the garden all that time and now have several areas where it makes extensive ground cover. After noticing it growing on the trunk of an *Abies* sp. one time at Mineral del Chico, Hidalgo, I decided to wedge a rosette on the trunk of my yew tree when I got home, making sure that it was steady until it had rooted. It has enjoyed an epiphytic life ever since. Encouraged by this, I put some more in the leaf base pockets on the trunk of my *Phoenix canariensis* and they have settled in rapidly there, too. Grown this way the plants seem to struggle during extended drought and benefit from an occasional splash of water, but otherwise I leave them well alone, wherever they are growing.

Echeveria elegans growing epiphytically at 3,000m altitude at El Chico, Hidalgo.

Echeveria albicans.

Echeveria albicans is a similar-looking species, though not as delicate, with fatter leaves and smaller rosettes. It is not anywhere near as hardy but, irritatingly, you sometimes see it offered for sale under the name of *Echeveria elegans*.

Echeveria rosea

Common name: none applied
Distribution: Mexico
Hardiness: H4/5

A variable plant with, typically, loose rosettes of glaucous-grey, lance-shaped leaves growing on the tips of an elongated sparsely branching stem, to 60cm or more. Flowering stems emerge from the side of a rosette, clothed in leafy bracts and ending in a congested spike of pale yellow flowers peering out from among colourful bracts. The grey is tinged with pink on the leaf margins and stems, this colour deepening as flowering commences until, when in full flower, the whole plant can look as if it has been dipped in pink, especially when grown in strong light. I have a rather striking form in which the leaves are glossy green and, rather than pink, the whole plant flushes with a strong oxide-red when flowering.

This has a wide distribution in Mexico, growing from Tamaulipas in the north all the way down the central and eastern states to Chiapas in the south, but almost exclusively as an epiphyte in a cloud forest environment. I have seen it growing on trees in a misty abies forest at around 3,000m in Oaxaca and also in Puebla on some damp, mossy rocks. This gives the best clues to how to grow it – perfectly drained medium, not too hot and a general dislike of completely drying out.

I have grown the regular form for some years, plants taking my worst three

The usual pink form of *Echeveria rosea*.

The unusual red form of *Echeveria rosea*.

Echeveria rosea growing epiphytically in habitat.

winters without issue. The red form I have only had for around eight years but it has been fine through my early and late cold winters. In an extended drought the plants can look a bit straggly as they lose leaves but soon recover in cooler, wetter weather. Flowering here is always in late winter or early spring, giving a most welcome hit of colour just when you need cheering up. They are easily propagated from basal rosette leaves or, better, the leafy bracts on the flowering stems.

Worth Considering

Echeveria agavoides

These are gorgeous, dumpy, chunky plants making fat rosettes that can be anything up to 40cm across. Coloration varies from plain greenish-grey to deep burgundy: sometimes leaf tips and margins are coloured most strongly, sometimes the whole plant is suffused with pigment. There is a particularly striking cultivar named 'Ebony' with pale grey-green leaves that contrast markedly with almost black leaf tips and edges but, really, any or all of them are exceptionally attractive things. In the open garden they struggle much below –5°C, especially if allowed to get wet or snow-covered. I lost some lovely plants in my early cold winter. Kept dry they will take –8°C quite happily. (H3/4)

Echeveria agavoides.

Echeveria agavoides with colourful leaf tips.

Echeveria secunda var. *glauca*

Superficially similar to *Echeveria elegans*, but not quite as elegant, these have wider, flatter rosettes of thinner but wider leaves in a pale blue-grey colour. Flower spikes have a stronger pink colour but flowers themselves are paler yellow. This species is extremely hardy, having taken my worst three winters unprotected. (H4/5)

Echeveria secunda var. *glauca.*

Echeveria × *imbricata*

This is another tough customer, similar to *Echeveria secunda* var. *glauca* (one of its parents) but with a denser rosette of thinner leaves. Flowers are nice and colourful with coral-red stems and buds, yellow when open. This one grows into large colonies of tightly packed plants, individual rosettes often being distorted and squashed when confined. It has been hardy here through my worst three winters. (H4/5)

Echeveria × *imbricata* at Crete Lodge.

Watch List

Echeveria chihuahuaensis

This is a fairly new plant for me outside, a veteran of just two winters as I write. It has taken my early cold winter with some damage but will almost certainly recover – which is more than can be said for a lot of succulents I have tried recently. It is an attractive species growing into a large rosette of grey leaves tinged red at the tips and leading edges. (H3/4?)

Echeveria chihuahuaensis.

Echeveria lilacina

This is a stunningly beautiful plant, growing into a broadly incurved, full rosette of palest lilac, almost white, leaves to 20cm across. I grew this outside on my vertical rock face for a number of years, where it took my late cold winter without a problem. But it was unable to cope with an extended drought and heatwave. I aim to replace it. (H3/4?)

Echeveria lilacina.

Echeveria prolifica

This is a cute thing, making massive colonies of small but chunky, powder-blue rosettes with offsets that are produced on long runners. I have been growing this for around five years and it has taken my early cold winter with only minimal damage. I think that, if kept dry, it would be fine down to around –8°C or maybe lower. (H4)

Echeveria purpusorum

This is a remarkable plant with the most gorgeously mottled leaves and quite unlike others in the genus. I kept this going for about five years in a rock crevice that kept it dry, taking my late cold winter just fine, only to lose it during an extended drought and heatwave. Apparently one problem with this is getting accurately named plants, as hybrids usually strongly resemble the species. I guess if it looks nice then it does not really matter. (H3)

Echeveria 'Perle von Nurnberg'

One of the hardier hybrids, this is a subtly coloured plant with tones of pink and pale mauve, forming a goblet-shaped rosette of dessert spoon-shaped leaves. Kept dry, plants have taken my late cold winter but unprotected they have been killed at –5°C. (H3/4)

Echeveria prolifica.

Graptopetalum

Family: Crassulaceae
Subfamily: Sempervivoideae
A genus of around twenty species found mainly in Mexico, one of which is described here.

Keystone Species

Graptopetalum paraguayense

Common name: ghost plant
Distribution: north-east Mexico
Hardiness: H5

Graptopetalum paraguayense is a highly attractive trailing shrub with lax stems that can reach 100cm or more in length but generally break long before then. The open rosettes are formed of leaves that are chunky, 6cm or more long, and a beautiful shade of grey that has just a hint of lilac colour giving the impression they are somehow made of mother of pearl. This changes in winter to a more solid blue-grey

Graptopetalum paraguayense.

and in summer, if stressed, to yellowish-grey. Rosettes are on the stem tips and, if grown in a vertical rock face or in a hanging basket, perform a kind of slow-motion abseil as the stems elongate. More shoots are made at the base of the plant and an established plant sporting various-sized stems really is a sight to behold. In spring branched inflorescences appear from each rosette carrying stems of white, five-petalled flowers, each with a red dot. Stems and, particularly, leaves are fragile and easily broken off. Each leaf will happily grow into a new plant, forming roots and little plantlets at the broken end. As and when it gets too scruffy, cut it back and let it all start again – plus you can root the stem tips and fallen leaves for new plants.

Graptopetalum paraguayense is one of the most widely grown plants across the world yet its origins are shrouded in mystery. First brought to the world's attention in New York in 1904, it was originally believed to come from Paraguay and named with that in mind. In reality it most likely originated from north-east Mexico but no one has actually found anything corresponding to the original in the wild to be certain of this. A similar but smaller plant was found in the state of Tamaulipas, described in 1979 as a subspecies of *Graptopetalum paraguayense* but that is as close as anyone has ever managed.

Leaving that aside, its elegantly soft appearance and the fragile nature of its stem both belie a rugged constitution. This is a very tough plant, known to withstand temperatures down to –15°C if covered, or –10°C in the open garden, even if covered in snow. I have grown this for over 30 years and still marvel at the improbability that this delicate beauty is so resilient. If you do not grow this plant, find someone who does and ask for a leaf. It could not be simpler.

Sedum

Family: Crassulaceae
Subfamily: Sempervivoideae

Sedum used to be a massive pan-global genus of succulents, containing over 600 species, but recently about half of them have been split off into several new genera. Their growth varies from small annuals to herbaceous and evergreen perennials and even large shrubs. With such a diverse and widespread group of plants it is inevitable that many are suitable to grow outside in the UK. I have tried just a few and look forward to trying many more, as and when I apply myself to sourcing material. Below are some of my favourites: some rare, some in every garden centre. A few of the more herbaceous types appear later under 'companion plants'.

Keystone Species

Petrosedum sediforme

Common name: pale stonecrop
Distribution: southern Europe to north Africa
Hardiness: H6

Petrosedum sediforme (syn. *Sedum sediforme*) is a chunky member of the needle sedums, superficially resembling the UK native *Petrosedum rupestre* (syn. *Sedum reflexum*) from which it differs by its thicker stems and fatter leaves that are densely clustered around the stem in a spiral. Overall it is a larger plant, growing to 20cm tall. Flowers are creamy coloured. In some ways it resembles the tender burro tail, *Sedum morganianum*, but smaller and, of course, a lot hardier.

Anyone going on holiday in southern Europe or the Mediterranean is bound to encounter this species as they walk along the coast. It varies quite a bit from population to population, as well as from country to country, with some forms barely distinguishable from *Petrosedum rupestre*. I have a particularly compact and chunky form that came from the Algarve, Portugal, quite near to Lagos. It has also naturalised in a few places in the UK, including a wonderful little nature reserve on Canvey Island called Canvey Wick. This was once an oil refinery, built but never used, abandoned to nature 40 years ago and now boasting a higher species count than anywhere else in the country.

I have grown this for many years, along with the similar *Petrosedum rupestre*, as trouble-free ground cover. *Petrosedum sediforme* has the edge when it comes to looks and general habit, being bigger and fatter it somehow looks more 'exotic', but *Petrosedum rupestre* has the edge when it comes to flowering, its bright yellow

Petrosedum sediforme growing in Mallorca.

Petrosedum sediforme originally from the Algarve.

Petrosedum rupestre.

Petrosedum rupestre softening the spikies with a yellow haze.

flowers held in a shimmering haze when caught by the breeze. Unlike *Petrosedum rupestre*, *Petrosedum sediforme* doesn't seem to self-seed, but the main drawback of both is having to snip off individual spent flower heads – if you pull at them they tend to come out with the plant still attached.

Sedum decumbens

Common name: none applied
Distribution: central/eastern Mexico
Hardiness: H3/4

Sedum decumbens is an evergreen perennial/sub-shrub that forms lax branching stems on the growing tips of which are cute, plump, apple-green, shiny rosettes of oval leaves. Small panicles of bright yellow, starry flowers appear in late winter. It is similar to *Sedum confusum*, which is slightly larger in all parts and tends to grow more upright.

I have seen both in their Mexican habitats. *Sedum decumbens* was growing in some fairly moist, shaded woodland in Veracruz near to Xalapa. *Sedum confusum* grew in a more open spot but still a relatively moist environment near Tepoztlán in Morelos, south of Mexico City. Both attract the eye with their glossy foliage, and individual shrubs attain a good size if left undisturbed.

Sedum decumbens.

Sedum decumbens in Veracruz.

Sedum confusum in Tepoztlán.

I have grown both for something like fifteen years. *Sedum decumbens* is the hardier of the two and made a spreading patch of bright green around 100cm across before my badgers decided to rip it to pieces one night. Cold and snow do not seem to bother it. *Sedum confusum* is a similar size and has been damaged in my worst and early cold winters but is still alive and recovering. It has also been trampled by the badgers so I can only guess the plants contain something they find attractive. They are not the only ones: I find both of these species to have immense value, their glossy/satin surfaced leaves brightening up areas of the usual glaucous grey that the xeric plants generally offer.

Sedum oregonense

Common name: cream stonecrop
Distribution: Oregon, USA
Hardiness: H6

Not to be confused with *Sedum oreganum*, *Sedum oregonense* makes sprawling ground cover with cute, incurving, glaucous, blue-grey leaves that have a distinct notched tip. Rosettes get to around 5cm across, offsetting freely to slowly make a mat. Heads of creamy white flowers are borne on pinky-red stems in spring.

Sedum oregonense.

This comes from just two mountain ranges in Oregon, where it favours open, sunny, rocky, gravel or scree slopes. In the garden it is best grown hard to keep it compact, so that the individual heads are bunched together. My plants have been growing for nearly 30 years, originally a present from plantsman Michael Prime who was growing a huge patch of it in his Lewisham garden. It has taken my three bad winters, extended periods of drought and wet, snow and ice, all without complaint. To me it looks far more exotic than its Oregon origins would suggest and these days it is usually found for sale in the alpine plant section of most garden centres.

It is ostensibly similar to *Sedum laxum*, from which it largely differs in the colour and structure of the flowers, but this is a species I have not grown yet so have no hands-on experience. One of many I have lined up to try in the near future.

Sedum palmeri

Common name: none applied
Distribution: north-east Mexico
Hardiness: H4/5

Sedum palmeri is an evergreen perennial with thin, brown, trailing stems topped with cute, chunky china-blue rosettes of oval leaves, some forms edged with pinkish-red. Bright yellow flowers are borne on branched stems in late winter and it is entirely possible to have a plant with the blue leaves, bright red edges and yellow flowers all happening at the same

Sedum palmeri.

time. As the stems elongate they arch over daintily, making the whole plant extremely attractive. It is variable in the size and shape of the leaves and also the degree to which the edges can turn red.

This comes from north-eastern Mexico. I have seen an especially dwarf form of this cascading out of a rock cut on a largely disused road south-east of Ciudad Victoria in Tamaulipas, growing alongside an assortment of xeric ferns, *Hechtia* spp. and a selection of agaves.

In the garden this is a colourful plant with an extended period of interest, especially with the early flowering and red flushing of leaves contrasting with the pale blue rosettes. I planted it originally in my vertical rock face with the idea it would dangle enticingly. An unexpected surprise was that, as the nearby plants grew bigger and restricted the space for the *Sedum*, the stems grew through and in between the leaves of its neighbours, making an interesting patchwork of colour and texture. I have grown it for twenty years with no damage.

Worth Considering

Phedimus spurius
(syn. *Sedum spurium*)

Another stalwart of garden centre alpine sections, this is an extremely common plant with several named cultivars, each of which brings something slightly different to the party. My favourite has to be 'Purple Carpet' (more accurately 'Purpurteppich') which starts the year deep maroon and just keeps getting darker and darker as light levels increase. Winter sees the little flat rosettes disappear down to smaller resting buds but it quickly gets going in

Phedimus spurius 'Purpurteppich'.

spring. Yes, it can be invasive in small places but to carpet extensive spaces under glaucous upright plants it is invaluable: errant stems are easily pulled out. (H6)

Sedum dasyphyllum

This is an extremely variable species that is found over a huge range from central Europe, the Mediterranean and into North Africa. Consistent features are tiny rosettes, crammed together like a head of glaucous-blue cauliflower, sometimes furry, with starry, white flowers. They are exceptionally drought-tolerant and best grown in rocky, gritty soil to keep them compact. Some forms are a little bigger, not all are hairy, but they are all as cute as anything. Plants and individual rosettes are quite brittle, falling apart if handled carelessly, but if it manages to remain intact, the sight of this tumbling over a wall edge, as though poured from a jug, is always welcome. I have grown this in my garden for fifteen years or more and it has taken my three bad winters just fine. (H5)

Sedum dasyphyllum.

Sedum praealtum

In many respects this is like a giant form of *Sedum confusum*, described above. Another Mexican species, this grows into a rather sprawling shrub to around 90cm high and

Sedum praealtum.

across, with weak stems that would like to remain erect but never quite seem to manage it. Satin green leaves and heads of yellow flowers complete the picture. I have grown this for over twenty years, during which time it has been damaged by each of my bad winters but always somehow managed to grow back again. I don't think it would take much lower temperatures than –8°C, unless it was covered to keep it dry. It is extremely easy to grow from cuttings, so I always keep a couple by way of insurance. (H3/4)

Sedum spathulifolium

This is a cute little stonecrop that can be seen for sale in garden centres the length and breadth of the UK, being one of the

Sedum spathulifolium 'Cape Blanco'.

top-selling rockery plants. It grows into dense mats of tiny rosettes, glaucous silver with a hint of purple. In nature it is found growing on rocky, open slopes of mountains along the western coast of the USA from California to British Columbia. The most familiar and popular form enjoys the full name of *Sedum spathulifolium* subsp. *pruinosum* 'Cape Blanco'. I love this plant. The hotter and drier it is the more the silver colour intensifies. There is another subspecies, subsp. *purpureum*, that is worthy of mention as the whole plant takes on reddish-purple hues. (H5)

Watch List

Sedum dendroideum

This is a shrubby sedum growing anything up to 2.5m high in nature. It is similar to *Sedum praealtum* but has a sturdier habit, with leaves that curl inwards. There is a cultivar named 'Colossus' that is nothing short of magnificent. Not only do the leaves curve inwards along their length, they curl outwards across the width, giving the whole thing a wonderful three-dimensional effect. Unfortunately 'Colossus' is not quite as hardy as *Sedum praealtum*; my outdoor plant was cut to the ground during my early cold winter having survived down to –4°C in a previous year. But it is really easy from cuttings, so keeping a potful frost-free under glass would pay dividends. (H3/4?)

Sedum dendroideum 'Colossus'.

Sedum oxypetalum

This is a shrubby plant, growing on a fat trunk to around 1.2m tall in the lava fields across central Mexico. It has narrow, flattened spoon-shaped leaves around 3cm long and the trunk develops the most beautiful peeling bark as it ages, swelling to around 10cm or more thick. In the dry season the leaves fall. If grown under glass and kept dry, this happens in winter, the plants leafing out again in spring. Starry white flowers are produced in summer at stem tips, which are said to be fragrant, although I have never noticed. This is a magnificent plant to grow, potted, as an easy-care bonsai and a truly magnificent plant in a cold glasshouse. Encouragingly, it has shown some hardiness. I know of a plant in central London that has survived down to –4°C when dormant. My own small plant has taken –3°C at the base of a tall conifer. I wonder just how low this will go? (H3/4?)

Sedum suaveolens

New to me, this is a beautiful ghostly pale silvery rosette that is more reminiscent of an echeveria than a sedum, spoiling the deception with its white, sedum-like and sweetly scented flowers. Offsets are made generously. I lost one of just two plants in my early cold winter so will regroup when I have more spares to find the right place but, for sure, it is a very interesting un-sedum-like sedum. (H3?)

Sedum × rubrotinctum

This looks like a stem threaded with red jelly beans, especially grown in full sun. It is not very hardy, only taking down to perhaps –3°C, but is extremely easy from cuttings, so can be used as a temporary filler to provide a splash of intense colour through the year. Of course, the bonus is that it will survive mild winters and get off to a quick start the next year. (H3)

Sedum × rubrotinctum.

Sempervivum

From the mountains of southern Europe to North Africa, this is a genus of around 50 species of small, rosette-forming, clustering succulents. They have been immensely popular for a very long period of time and around 3,000 cultivars have now been named. Given that a single cultivar can vary a lot in appearance according to the time of year and growing conditions, it is fairly pointless making any recommendations. The best thing to do is pop into any garden centre and choose the ones you like the look of, ignoring the label. Recently the guys at Surreal Succulents, near Penzance, have caused something of a stir in the horticultural world by hybridising *Sempervivum* with *Aeonium*, the result being called × *Semponium*. Various cultivars have been selected and named and it is all very exciting as these plants should, in theory, have inherited hardiness from *Sempervivum* but a range of strong colour and form from *Aeonium*. Standout cultivars for me are the pink-red blush × *Semponium* 'Sienna' and the deep maroon 'Destiny'. Early trials, however, have been rather disappointing with the

Sempervivum are just so easy, accommodating and pretty.

Beautiful cobweb rosettes of a *Sempervivum arachnoideum* form.

Umbilicus rupestris in Mallorca.

Umbilicus oppositifolius.

anticipated hardiness not making itself apparent.

Umbilicus

A genus of interesting little plants found on most continents of which two species are of particular interest. *Umbilicus rupestris* is native to the UK and all the way east to North Africa, via the Mediterranean. Fresh green, peltate leaves arise from a tuberous rootstock. These are fleshy, circular with a scalloped edge and central dimple, leading to the common name of navelwort. After a couple of years it produces an upright spike of slightly nodding green flowers and sets copious seed, which then spread around far and wide, thus completing the life cycle. It favours moist, shady rock crevices and in a mild, wet climate, such as the south-west of Ireland, it remains evergreen with leaves up to 8cm in diameter. I have also seen it growing in Sicily and Mallorca, where it dies back to the tuber in late spring with leaves, when they appear, to just 2–3cm across. Grown hard in full sun they often take on red tints, some plants I saw on a wall in Mallorca being a gorgeous plum-purple. I started off with one plant, a gift from a friend in Ireland from his wall, and managed to keep it going until it flowered, which is not the easiest of tasks as it hates being disturbed. Having flowered and set seed the plants have now spread to wherever they want to be and if that happens to be somewhere they are not wanted, they are easily removed.

Umbilicus oppositifolius was called *Chiastophyllum oppositifolium* when I first bought it but is now considered to be an *Umbillicus*. This is from the Caucasus Mountains and, like its cousin, likes moist, shady rock crevices but will also grow in drier positions in full sun. Apple-green, succulent leaves are oval with a toothed margin and the flowers are remarkable tumbling cascades of yellow catkins. This will also self-seed around but without quite the same enthusiasm. Both species are invaluable for keeping the succulent theme going in shadier positions and are reliably hardy in the UK.

MESEMBS AND SPURGES

Aloinopsis

A genus of leaf succulent mesembs (contraction of Mesembryanthemaceae, former name for the family Aizoaceae) growing into a congested bun of densely packed, spoon-shaped leaves that are highly textured with knobbles, bumps and warts. Flowers are produced in early spring. A few show good hardiness if kept completely dry over winter and only watered sparingly in summer, of which *Aloinopsis spathulata* is the toughest, said to take down to a ridiculous –20°C if dry. Unusually, it has pink flowers. Leaves are spatulate (broad across the top, tapering to the base), blue-grey and covered in small warts. I have tried this a few times but never managed to get it established for long. I grew some from seed around twenty years ago, planted out three, but was careless about it and lost them in the first winter. My next attempt barely made it back home from the nursery as, upon taking it out of the pot to plant it, the whole thing fell apart due to vine weevil. I lost my third go with it in the open garden after a particularly wet autumn. Never easily discouraged, I am trying it again somewhere with more cover. (H5/6?)

Aloinopsis spathulata.

Titanopsis calcarea is one of another genus of mesembs that is similar in overall appearance to *Aloinopsis*. The leaves are thicker and blunter at the tips and have a knobbly and warty texture reminiscent of a particularly nasty skin complaint. Flowers are orange-yellow daisies, produced in winter. I grew one of these in my vertical rock wall for something like eight years until an extended drought and heatwave saw it off. Plants are hardy to maybe –8°C if dry. (H4/5)

Titanopsis calcarea.

Delosperma

Family: Aizoaceae
Subfamily: Ruschioideae

This is a large genus of 170 odd species of, mainly, trailing, shrubby, succulent plants often with showy daisy-like flowers. Several species are found for sale in the alpine sections of garden centres and many new hybrids are being bred to increase the range of flower colour. They are one of the larger genera of mesembs.

Keystone Species

Delosperma cooperi

Common name: hardy ice plant, *Skaapvygie*
Distribution: South Africa, Free State and Lesotho
Hardiness: H5/6

Delosperma cooperi grows into dense mats of narrow, fleshy, finger-like fresh green leaves around 8cm long bearing, from late spring until early autumn, stunning 5cm-diameter, magenta, daisy-like flowers with no central white on them at all, just the contrasting yellow stamens. If allowed, it will spread to 2m across or more. It is found on dry grassland at high altitude and is extremely tolerant of different growing conditions, as long as the drainage is good.

In the garden it can take over somewhat if left unchecked – I find plants are best trimmed back to within a couple of centimetres of older, woody stems every spring, using pruned material for cuttings. That way plants can be replaced every few years when they start getting too woody. I have grown this for 25 years and it has never been worried by bad weather. The flowers are gorgeous, a bright splash of colour, produced sporadically over a few months. You never get the same intense, eye-watering blast of colour that you do with, say, *Lampranthus roseus*, but it has a longer season and extra hardiness to compensate.

I have seen plants labelled *Delosperma cooperi* 'Compacta' that have stubby

Delosperma cooperi.

leaves, washed out purple flowers, with a white middle that are not produced in great numbers. I do not know what it actually is but it has nothing to do with *Delosperma cooperi* and is a much inferior plant.

Delosperma 'Compacta'.

Delosperma nubigenum

Common name: *Lesothovygie*
Distribution: South Africa, Free State and Lesotho.
Hardiness: H5/6

Delopserma nubigenum is an absolute gem of a plant. It grows into a trailing, dense mat of glossy, apple-green leaves the size of babies' fingertips. In early summer these are covered by 3cm-wide, pure golden, silky-textured, yellow daisy-like flowers and followed in hot, dry years by the foliage turning red when stressed.

Delosperma nubigenum.

In some years you can have all colours happening at once. It does not reach more than 2cm high but can still make a patch 90cm or more.

This is another plant from high in the Drakensberg, which, when actively growing, will tolerate moderately damp soils as long as they drain well. It is still best kept drier in winter, though. It is hard to overstate how useful this plant is: in full sun on well-drained, poor soil this has few peers. It is reliably hardy – one of the absolute hardiest mesembs taking down to –15°C – and so very pretty in flower. It makes very low ground cover so does not really interfere with the plants growing above it. It is very easy to control if it starts to spread a bit too much – just snip it off near where you want its limits.

There is a cultivar named 'Basutoland' but, having grown both, I cannot actually see how this differs from the type. The species *Delosperma congestum* has a pale centre to the yellow flowers but it is otherwise similar, so I have not tried it.

Worth Considering

Delosperma sphalmanthoides

An absolutely stunning plant that has the narrowest, grey-green, finger-like leaves packed into a dense dome giving the overall impression of a sea anemone. Magenta flowers cover the mound completely, just adding to the desirability. I have found this infuriatingly impossible to grow in the open garden but will keep trying until I succeed: I believe keeping it totally dry over winter and fairly dry the rest of the time is the key. I have just the spot in mind. (H4/5?)

Delosperma sutherlandii

This is a smaller plant than the previous species, making compact growth from a deep tuberous rootstock. Flowers are purple with a pale centre to 2cm across. It is another high-altitude species and unlikely to outgrow a pocket in a rock garden or interfere with its neighbours. I lost mine during an especially wet winter. (H4/5?)

Delosperma sutherlandii.

Delosperma 'Jewel of the Desert' Series

These are a new group of hybrids that are appearing in garden centres; they have a much expanded range of flower colour. Names like 'Garnet', 'Peridot', 'Grenade', 'Topaz' all have eye-catching colours. I do not know the parentage – I imagine the exact formula is still a trade secret as they are subject to Plant Breeders' Rights – but the ones I grow appear to have affinity with *Delosperma nubigenum* in their growth habit and tolerance of moister soils. (H4/5?)

Drosanthemum

Another genus of trailing mesembs distinguished from *Delosperma* by having sparkling, jewel-like leaves. It is almost worth buying a hand lens just to see these crystalline leaves in close-up. Some species have the most incredibly showy flowers but are, generally, not as hardy as the *Delosperma*. Nevertheless many species are worth growing, one of my favourites being *Drosanthemum micans*. The leaves are glistening grey-green and the flowers pure shining gold with deep purple tips – an absolutely stunning combination. They can get a bit straggly so benefit from being clipped over after flowering – at which point you have plenty of insurance cutting material. Hardy to around –5°C. (H3/4)

Somewhat hardier, taking down to –10°C if dry, *Drosanthemum hispidum* is a bit tidier, forming a small mound of greener leaves and pale purple flowers. (H4/5)

Perhaps the brightest of all, *Drosanthemum speciosum* has vivid scarlet flowers and is hardy to –8°C if dry. (H4/5)

All are easy from seed and cuttings.

Drosanthemum micans.

Cautionary Advice

There are two trailing/creeping, mat-forming succulent plants that are commonly offered for sale. I have grown both and regretted it.

Carpobrotus edulis grows by the hectare on coastal cliffs in milder spots throughout the UK, but also anywhere it has been introduced: where the climate allows, it has become a serious ecological problem. In Mallorca, for example, there are various initiatives to eradicate it; last time I was there I saw huge bundles on the sides of beaches along the north coast, waiting to be collected. It is a rampaging thug of a plant with no place in a domestic garden. The leaves are quite nice but there are too many of them on the several metres of stem that it produces annually and nowhere near enough flowers. Grow it in a hanging basket if you must but never let it escape.

The second is actually even worse. *Aptenia cordifolia* has cute little heart-shaped leaves and even cuter little raspberry-red flowers. Before I knew any better, I planted both together in a spot in my garden and the *Aptenia* actually swamped and outgrew the *Carpobrotus*. Both will harbour slugs and snails in their weave of old stems. Just don't.

ABOVE RIGHT: *Carpobrotus edulis* on Lizard Point, Cornwall.

RIGHT: The glossy leaves and bright foliage of the *Aptenia cordifolia*.

Euphorbia

Euphorbia is one of the giant genera in the plant kingdom, with over 2,000 recognised species that are spread across every continent. They are universally known by the common name of 'spurge' and can be anything from annuals to tall trees. One thing that unites them is their unusual flower structure, a strange thing with various floral parts fused together called a cyathium that is not found in any other plant group and is, I suspect, the principal reason why the genus has so far been spared the taxonomist's knife. Many have evolved succulence, yet only a few show much cold-hardiness. Other, non-succulent, drought-tolerant species are included in the companion plant section.

By some measure the coldest hardy succulent species is *Euphorbia clavarioides* var. *truncata*. This is a weird plant from South Africa, the range of which in the wild includes some high Drakensberg peaks in Lesotho. It grows from a turnip-like taproot growing many finger-sized, knobbly stems and, with age, inexorably moulding itself around everything in its path. Zoom in close and the stems and, particularly, the flowers are fascinating in the detail. Take a few steps back and marvel at the spreading mound. Plants have been documented to take temperatures below –15°C with no harm at all when dormant. Mine had grown happily for over twenty years, taking my worst and late cold winter with no problems. My early cold winter, however, caused the upper stems to go brown – clearly it is not happy taking ice and snow when actively growing.(H5)

From the Atlas Mountains in Morocco, *Euphorbia resinifera* is almost like a giant version of the above, in habitat growing into huge, domed mats, up to 2m across, of four-angled stems armed with the occasional but sharp spine. I have grown this several times but never managed to keep it going for longer than about four years. If dry this can take down to –8°C but it dislikes getting wet and absolutely hates snow and ice. (H4)

Euphorbia royleana is a densely branched, shrubby species growing up to 5m tall from across the Himalayas at altitudes up to 2,000m. It has upright five- or six-angled branches and quite large, fleshy leaves that are produced during wetter and warmer periods but shed when temperatures start to fall. This is said to be hardy down to –8°C but, as yet, I have never managed to keep it for more than two years. I must keep trying as the prospect of a large, shrubby, succulent *Euphorbia* is very exciting so, with that in mind, I keep a large mother plant in my

Euphorbia clavarioides var. *truncata*.

Euphorbia resinifera in Naples' Botanic Garden.

Euphorbia royleana.

Euphorbia stellispina in my greenhouse.

Euphorbia caerulescens.

frost-free greenhouse from which I take root cuttings. Certainly in the wetter, milder south-west, adventurous gardeners should be experimenting with it. (H3/4?)

I grew *Euphorbia caerulescens* for a few years before my bad winter saw it off. It is like a larger, not as densely growing but more brutal version of *Euphorbia resinifera*. I would say it would be safe down to –5°C if dry. (H3/4?)

One more species has recently come to my notice. Veteran plant pioneer Peter Jenkins has had a potted specimen of *Euphorbia stellispina* outside in his garden for a couple of winters, taking down to –5°C in a sheltered position against the house wall. This needs further investigation. (H3/4?)

Faucaria

A South African genus of around 30 species of leaf succulents with interesting form and markings. I have grown a few and find them to be remarkably hardy, growing in crevices in my vertical rock face. Two species have grown here for many years.

Faucaria tigrina has fat, roughly triangular leaves that are arranged in opposite pairs, the edges of the leaves displaying extended marginal teeth – the look reminiscent of the jaws belonging to some type of fearsome alien predator, albeit a rather small one. The surfaces of the leaves are covered in dots, spots and warty markings and, if stressed during hot, sunny weather, can turn red or maroon. During winter a bright yellow, daisy-like flower is produced that seems impervious to inclement weather, some years flowering happily in the snow. *Faucaria tuberculosa* is similar, believed by some authorities to be a subspecies of *Faucaria tigrina*, but leaf surfaces, especially the inner surface, are covered to a

Faucaria tigrina.

Faucaria tigrina flowering in the snow.

Faucaria tuberculosa.

greater or lesser extent in exaggerated pale, warty bumps or tubercles. The marginal teeth are generally smaller but the yellow winter daisy-like flower is much the same. These have both taken –8°C in my garden during my bad winters, with the caveat that they are tucked neatly into rock crevices. (H4)

Hereroa

Hereroa glenensis

I bought this as *Bergeranthus glenensis* but believe it is now more correctly *Hereroa*. I also have other species that are similar, one of which might still be a *Bergeranthus* – *Bergeranthus scapiger*. In any case they are much the same: clusters of dark green, angular, finger-like leaves, around 7cm long, are held upright from a creeping rootstock. Yellow, daisy flowers are produced on short stems, opening in the afternoon. *Hereroa glenensis* has the added interest of a reddish-brown reverse to the flower. These are extremely hardy having sailed through my bad winters. I understand that they will take temperatures down to –15°C or lower. They are perfect little

Hereroa glenensis.

succulent fillers for tight crevices and corners in a rock garden, the rootstock expanding to tightly pack the crevice so that the plant seems to be growing straight out of the rocks. (H5/6)

Lampranthus

A South African genus of around 230 species of trailing or shrubby leaf succulents, some of which have the most impossibly bright, almost fluorescent, flowers of any plant in the succulent garden. I mention three species here, but there are now so many hybrids and cultivars available that the best way to choose one is to visit a nursery when they are in flower and pick the colour you like. They all require the same sort of care: a position in bright sunlight, perfect drainage and, ideally, a wall or rock to tumble over. None are reliably hardy in my climate, anything below –5°C is going to kill them off in the open garden, but they are easily raised from cuttings. As they are best trimmed after flowering there should never be any shortage of material. Of course, plants are fine in milder winters: I would say, on average, that I lose my *Lampranthus* once every four years or so. A reasonable question would be why go to the effort when there are hardier plants like *Delosperma cooperi* and *Delosperma nubigena*? It is all about floral fire-power, really. *Lampranthus* will flower for maybe a month but in that time the entire plant is hidden beneath a retina-scorching display of dense flowers.

Lampranthus aureus, orange form, providing early colour.

The intensity of the colour is such that it is almost impossible to capture in pictures.

My absolute favourite is *Lampranthus aureus.* This species has relatively chunky leaves and grows into a 60cm-diameter mound, as opposed to spreading horizontally indefinitely like some species do. The daisy-like flowers are huge, 8cm across and, for my money, the best in the genus. Two sizzling colour forms exist – the more common of the two is a startlingly bright orange, the other a glossy dandelion yellow, which I have not actually had for a few years now. For me they flower from April into May. *Lampranthus roseus* has smaller, grey-green leaves and

Lampranthus roseus magenta form.

Lampranthus spectabilis.

Lampranthus brownii.

Lampranthus roseus red form.

initially grows into a mound, soon spilling over and spreading. The flowers are 5cm-diameter beacons, the colour varying from a completely outrageous, almost fluorescent, cerise to a traffic-stopping magenta. *Lampranthus spectabilis* has slightly larger flowers that are a less intense pale magenta into lilac. (all H3/4)

One species I have grown that is a good deal hardier, down to –8°C if dry, is *Lampranthus brownii*. The stems are wiry and woody and the plants overall are not as neat. The flowers are smaller and a deep copper red. (H4)

I have another one that has taken my early cold winter without damage but I do not know which it is. It grows into a shrubby mound, not spreading, with reddish stems clothed in small, pointed, grey leaves and small, lemon-yellow flowers.

CHAPTER 8

COMPANION PLANTS

It would be impossible to attempt to list or describe all the plants that make suitable companion plants for the selection of cacti and succulents that I have chosen – there are just so many. Most books on Mediterranean-style or alpine plant gardening would be far more comprehensive than anything I could manage to include here, the main thrust of this book being cacti and succulent plants.

To that end, I do not list 'Keystone Species', rather just loosely discuss the plants I enjoy growing that fit into my concept of the dry exotic garden. They might be reliable, hardy garden plants offered for sale in any garden centre. Some might be rather unusual and will need searching for but are worth trying. One or two might never become available but are interesting to discuss. As I progress towards the end, the detail becomes markedly reduced as there are just so many herbaceous plants, bulbs and ferns, so mostly I just mention the name and why I have included them.

A notable absence: there is no mention of any of the *Protea* family, a group of plants that I have no experience of growing. I do, however, have plenty of experience of killing them as they apparently object to something in my soil and die very quickly. It is such a shame, as there are plenty of interesting and attractive *Protea*, *Leucodendron* and *Banksia* species eminently suited to growing in a xeric garden. Just not here. I only want to include plants I am familiar with first-hand.

PALMS AND CYCADS

Palms form the main group of plants that I would associate with an exotic-style garden. There is nothing quite like a palm to conjure up visions of exotic holidays. I would, however, comment that not all of the hardier palm species fall under the remit of this book, as they grow so much better in wetter climates unless given regular, deep irrigation. I have also included some cycads and a couple of other non-palms, as it was as good a place as any to fit them in!

Brahea

Brahea armata

A beautiful silver-grey fan palm that slowly grows a fat trunk. In my drier region this grows perfectly well year-round with no cover but I understand that in wetter climates it is prudent to erect a shelter of some sort over them in winter to prevent them getting too wet. In summer that is not a problem and plants grow more quickly in reasonable rainfall areas. Mine

Brahea armata.

Closer look at foliage and brown tomentum on petioles of *Brahea armata.*

OPPOSITE PAGE: Stunning flowers of *Puya alpestris* at East Ruston Old Vicarage.

has been in the ground since 1996 and has slowly reached the point where it has a 1.8m-diameter crown, is over 2m high, with 90cm of clear trunk. (H5)

Brahea edulis

This is a personal favourite of mine, having the largest crown and leaves of any fan palm in my garden and it certainly looks as though it would not be out of place in Singapore or other warm, tropical places. It manages to tick along quietly, putting up with my low rainfall levels but noticeably putting on growth spurts after heavy downpours during the warmer months and, I feel, would be much happier in a wetter climate. (H5)

The same applies to *Brahea brandegeei*, which is particularly glacial in its growth rate, and *Brahea* 'Super Silver', which I also grow. A final mention goes to the super rare *Brahea moorei*, my plant being a treasured gift from my friend and unstintingly generous palm-grower in the Algarve, the late Charlie Wychgel. I have only had this planted out for a couple of years, set into a very calcareous substrate of rubble and crumbled mortar on a south-west facing slope that gets rather hot in summer. So far it is looking very promising, having made a little new growth each year. It has taken my early cold winter with no ill-effects but, really, should be hardy: I have seen it growing in north-east Mexico in the same woodland as *Chamaedorea radicalis*, which is a hardy plant for me.

Brahea edulis.

Butia odorata.

Butia

Butia odorata

These plants used to go by the name *Butia capitata* but when someone had a closer look at the paperwork it was realised that we should have been calling them *Butia odorata* all along. Confusingly, the name *Butia capitata* is the valid name for a completely different plant but, mercifully, that is rare in cultivation. As usual, name changes take a generation or two to become adopted in the nursery trade and plants are offered for sale under both names. With its grey-green, recurved leaves and, when older, beautifully scarred trunk, this palm has the right look to accompany succulent plants and will coast along slowly with no irrigation. They grow more quickly into much bigger, happier and prettier plants, however, anywhere west of here. The same applies to *Butia eriospatha*. (H5)

Chamaedorea

Chamaedorea radicalis

This is an unfeasibly cute little feather palm from north-east Mexico that grows as a solitary plant that rarely exceeds 70cm tall, graced with ranks of feathery, fresh green leaves. It flowers from a stem produced near the base: strange knobbly rat-tails with male and female flowers on different plants but readily setting bright red fruit if both sexes are grown together. Seed is usually viable. In the wild I have seen them growing on rocky slopes in the lightly shady understorey of pine and oak forest in pockets of organic soil. In my garden they are perfectly happy in the driest shade, disliking too much direct sunshine. For such a delicate-looking plant it is quite hardy; I know of plants that were defoliated at –12°C but have recovered. What they do not like is snow – the weight of snow on their leaves causes them to fold over and break, which rather spoils them. Sadly, they are very slow-growing and it takes a couple of seasons for them to get back their good looks. There is a trunking form that grows 3m or more on a slender stem that seems to be more tender. (H5)

Chamaedorea radicalis in woodland habitat, Tamaulipas.

Chamaerops

Chamaerops humilis

This rugged fan palm is just about the perfect plant to accompany succulents. Not only do the tough leaves with their spiny stalks look just right but, being a native of the Mediterranean, it is heat-loving and supremely drought-tolerant. In the wild they are often to be found growing wedged into tiny rock crevices in limestone cliffs. In time it grows into a many-stemmed, suckering, large shrub: after 25 years my oldest plant is now 4m high with six main stems. I remove any further suckers to prevent them from growing and, each year, assiduously trim off the oldest leaves so that, slowly, the leaning trunks are increasingly exposed, which also allows light to pass through to the garden behind. Apart from when I first planted it, I have never, ever given this supplementary watering. Surprisingly showy yellow flowers and, if they form, plump mahogany fruits are colourful bonuses. These are tough, hardy plants, generally taking down to –10°C before showing any damage. Even if defoliated at that temperature or lower, they will usually regrow from underground. When I bought my first one it had an issue with the fungal disease black spot, a common problem with this palm. I spent a couple of years trimming it heavily and regularly spraying with a systemic fungicide and now it is no longer a concern.

Trimmed stems of my *Chamaerops humilis*.

Chamaerops humilis is extremely variable and, historically, many 'unofficial' varieties have been named, depending on the habit of the palm, colour of its fruit and so on. Officially there is one recognised subspecies and a handful of more generally accepted cultivar names. *Chamaerops humilis* var. *argentea* was always known in palm-growing circles as *Chamaerops humilis* 'Cerifera' and that is the name that is still largely used, no matter what we are told. The leaves are pale blue and have a white, waxy coating that gives them a glaucous quality that is extremely attractive. Leaf stalks are generally armed with particularly vicious black thorns that provide a dramatic

Chamaerops humilis in habitat, Sicily.

Magnificent *Chamaerops humilis* in Tenerife.

Chamaerops humilis var. *argentea*.

Chamaerops humilis 'Vulcano'.

Chamaerops humilis fruits can be an attractive feature.

contrast to the pale leaves. This form comes from the Atlas Mountains in Morocco, rather than the Mediterranean, and is a couple of degrees hardier to cold than the type. *Chamaerops humilis* 'Vulcano' is a dwarf cultivar, bred by a Sicilian nursery, that has smaller, stiffer, less-divided leaves, generally with a strongly silver reverse and shorter, nearly thornless petioles. The plants originated from a small population on the Aeolian Island of Vulcano in the 1970s and a strict breeding programme is needed to keep these dwarf characteristics going, as a decent proportion of seed-raised plants lose them. Some of the plants discarded from the selection process turned out to be quite uniform, compact in size and resembling a scaled-down version of the type with quite deeply cut leaves but keeping the strong silver reverse. These have been marketed under the cultivar name 'Etna Star'. The cultivar name 'Vulcano' has been legally protected and cannot be used without royalty payments to the developing nursery; however, plants raised from seed taken from 'Vulcano' plants by other nurseries that show the same dwarf characteristics are now available under the cultivar name 'Compacta'. All are H5.

Cordyline

Cordyline australis

As one of the absolute stalwarts of the exotic garden, this is too familiar to need

Cordyline australis on the seafront at Eastbourne.

describing. I mention it in passing as it will tolerate dry growing conditions and, at a pinch, can do the same job visually as a trunking nolina with just a slightly different vibe, especially in a milder, wetter climate, where some nolinas might struggle. Plus, of course, the annual show of highly scented flowers is a bonus, the ensuing white berries providing a useful source of food for birds in winter. Temperatures of –10°C or lower can cut a plant to the ground but they will, in all probability, grow back even if it takes a while. I know of one plant that took eight years to regenerate, which is quite a remarkable feat. (H4/5)

Cycas

Cycas revoluta

The most frequently encountered cycad, *Cycas revoluta* is commonly seen for sale as a houseplant; its tough, leathery, glossy, deep-green, feather-shaped leaves being eminently tolerant of dry air indoors. These cycads, however, enjoy life outside and are surprisingly hardy to cold, taking down to –10°C unprotected; even lower with protection. In fact, they are rather difficult to kill but often equally difficult to please, taking a sabbatical year or four between flushes of new growth, by which time the existing leaves look a bit scruffy. A position in full sun helps considerably. When clearing a difficult and heavily shaded part of my garden recently, I discovered to my horror a forgotten *Cycas revoluta* that had been languishing in its leafless state for, I would estimate, eight years, possibly ten. To my surprise it did not look dead, so I transplanted it to an open, sunny site and, to my bigger surprise, the following year it flushed new growth. Eventually plants form a stout trunk and, on large specimens, it is common to see smaller crowns appearing from pretty much anywhere along the main trunk.

Cycas panzhihuaensis

This is another similarly hardy plant with attractive sea-green leaves and a curious spongy covering on the top of the caudex. New growth flushes are earlier in the year and also perhaps more reliable but, again, a position in full sun is best. Protection against molluscs is essential as the emerging leaves are completely irresistible to them.

Both of these sit well in the look of a dry garden but, ideally, should get periodically watered when flushing new growth. Both H4/5.

Cycas revoluta in Rob Stacewicz's garden.

New growth on my *Cycas revoluta* after a decade's sabbatical.

Cycas panzhihuaensis in James Knock's garden.

Jubaea

Jubaea chilensis

This is the supreme champion of feather-leafed palms for the dry garden. It is restricted to just a few small areas in central Chile, where it can be seen growing alongside cacti. The climate here is similar to that of the Mediterranean: hot, dry summers and cooler, wetter winters, with some areas seeing under 300mm of rainfall annually. In their habitat the palms live to a great age and can reach 25m or more in height on an immensely fat trunk that appears to have been made out of poured concrete. Stiff, feather-shaped leaves can reach 3m long on mature plants and, for such a lump of a plant, arch delicately to give a full, globe-shaped crown. The trunks of elderly palms, around 100 years old or so, taper inwards and resemble wine bottles. There is a commonly seen variant that holds its older leaves upright, the crown resembling a shaving brush, known ingloriously as the 'stiffy' form. There is an especially choice and desirable form with

Jubaea chilensis.

silvery-blue leaves that seems particularly prevalent in the region of the Italian Lakes.

In cultivation *Jubaea chilensis* is rather slow-growing but incredibly handsome at all ages. Young plants are squat with short, stiff leaves: teenaged plants are also squat with short, stiff leaves, just a little bigger. Once they reach about 25 years old they seem to finish expanding at the base and start to slowly make their way upwards. I have a specimen that I planted twenty years ago as a seventeen-year-old plant and it now has a trunk that is taller than me, although it has still retained all its leaf bases, giving it the appearance of a rather large pineapple. Of all the palm species I grow, something like 30 taxa, this is my absolute favourite, being perfectly in tune with my climate. It has never been damaged by snow or cold, never complained during extended droughts or heatwaves, just steadily gets bigger and better. The leaves are tough and last for many years before getting tatty, so I only generally need to remove those that get in the way; as a result the crown gets fuller and more impressive with age. (H5)

Jubaea chilensis 'stiffy' form in Atif Amin's Hampshire garden.

Jubaea chilensis silver form near Lake Maggiore, Italy.

Macrozamia

Macrozamia communis

If any macrozamia could be described as having any kind of track record for growing outdoors in the UK, this would be the species as, although by no means commonly grown, there are specimens dotted about the UK here and there. These are stupendously beautiful plants that form a short, stout caudex from which erupt fine, leathery, feather-shaped leaves, up to 2m long. The leaflets are arranged in a V-shape around the midrib at first but levelling out to a flat plane about halfway along. In nature these would receive somewhere between 1,200mm and 1,500mm of rainfall a year, so they appreciate a good soak occasionally in drier conditions. Unprotected the leaves become damaged and unsightly at around –5°C but, unlike *Cycas revoluta*, are replaced fairly consistently each year, both as a main flush of growth as the weather warms up but also steadily pushed from the middle throughout the year. With a fair amount of winter protection plants have been documented as recovering from as

Macrozamia communis in Darren Hobbs' Newport garden.

Macrozamia communis in Tenerife.

Macrozamia communis coning in Darren Hobbs' garden.

low as –10°C, which puts this species within reach of many gardeners. I know of a couple of long-established plants in Cornwall and Wales that have produced flowering cones.

Macrozamia moorei

With age this grows a massive trunk which, in its habitat, can attain 8m. As a young, trunkless plant it can be distinguished from *Macrozamia communis* by the leaves, which are held in a deep V-shape throughout their length. This has similar hardiness to *Macrozamia communis*, needs less water to grow well but more heat, so is only really viable in heat traps in hotter regions of the country. (H4)

Macrozamia moorei in a garden on Ischia, Italy.

Various other cycads have been tried here and there but are by no means widespread. Plants are not cheap to buy, making cold-hardiness trials an expensive risk. The following are the few I currently grow and I am allowing myself to become cautiously excited. I have grown *Lepidozamia peroffskyana* outside for eight years, during which time it has taken both my late and early cold winters with some protection. Leaf damage shows at around –3°C but complete defoliation at –6°C was followed by a new flush of growth the following spring. I grew *Encephalartos lehmannii* for

Encephalartos lehmannii.

Encephalartos lehmannii new growth after –5°C.

Encephalartos lehmannii in Madeira.

several years with flushes of new growth reliably appearing in spring after defoliation at –5°C. Then, as luck would have it, I was persuaded to dig it up and pot it immediately before my worst winter. At that point I was reluctant to protect plants and I am not convinced it would have survived that winter without any protection. I have since replanted it in a better position, together with *Encephalartos friderici-guilielmi* and *Dioon edule*, and all three have shown no damage at –5°C, producing new growth the following spring. My friend Darren Hobbs in Newport has grown *Ceratozamia robusta* for a decade or more and his plant has taken –8°C with protection, surprisingly staying green when unwrapped in spring.

Phoenix

Phoenix canariensis

The most familiar of the genus, available to buy from most florists, garden centres and DIY sheds throughout the land. It is not really a long-term prospect for most of the UK, being damaged badly around –8°C and usually killed stone dead at much below –10°C. They are also huge. Crown spread can be as great as 9m (I measured one of mine). If you have the space and the climate, however, they are spectacular, possessing that rare combination of immense size and elegance. I had three growing in my garden, planted in the late 1990s, and they were just fine until the winter of 2009/10, which killed one outright and defoliated the other two. The next winter killed off a second, leaving one remaining that has now rallied and regained its former looks, albeit with a constriction of the trunk to mark its experience. The other issue for me is that, although tolerating my dry growing conditions, they much prefer a wetter climate than here and grow so much better further west. (H4)

Phoenix canariensis on Southend seafront.

Phoenix loureiroi

I acquired this rarity completely by accident, as it came with the label *Phoenix acaulis*, which it clearly is not. A little research confirmed the ID. It is an elegant, dwarf-growing *Phoenix* with enchanting blue-green leaves that are very spiky at the base. It forms a short trunk and flowers as a small plant. The big surprise has been its cold-tolerance, coming through my worst winter with no damage. I had originally planted it in a completely inappropriate place and recently moved it to somewhere sunny where it seems a lot happier. (H4/5?)

Phoenix theophrasti

If ever a palm had 'the look' that is perfect for the dry garden it is this species. Those arching, spiky, silver leaves are just so redolent of desert oases, camels, Bedouin tents and Berber adobe buildings. Recent studies have shown this to be one of the parents of the commercial date palms, hybridised way back with *Phoenix dactylifera* from the Middle East. Unlike the commercial date palm, *Phoenix theophrasti* will happily grow in the UK, albeit slowly, its heat requirement apparently being low enough. It is also cold-hardy enough to withstand low temperatures in many parts of these isles. Palm pioneer Peter Jenkins has the champion UK specimen that has been growing in his front garden in Feltham near Heathrow Airport for 35 or more years: an absolutely peerless and majestic plant with 2m of clear trunk, the canopy a

Phoenix loureiroi.

The awe-inspiring *Phoenix theophrasti* in Peter Jenkins' garden near Heathrow.

Phoenix theophrasti detail of the lethal yellow thorns.

spreading cascade of silver leaves, making it quite unlike any other hardy palm it is possible to grow.

My own plant withstood my worst winter in 2009/10 with no damage, quite remarkable when compared to *Phoenix canariensis*. Of course I rewarded such a stoic individual by moving it to a 'better position' in the garden, whereupon it up and died. Finding a replacement was not easy but I now have one growing in the most suitable place I can find – its head in the sun, its feet in rubble, under which a spring nearly makes it to the surface. I am hoping the roots will tap into this, which should help it along in my dry summers. (H5)

Trachycarpus fortunei.

Trachycarpus wagnerianus.

Trachycarpus princeps, or possibly a hybrid.

Trachycarpus

No book on the subject of exotic gardening in the UK would be complete if it did not mention *Trachycarpus* palms, yet I am only going to mention them in passing. For certain, *Trachycarpus fortunei* is one of the most cold-hardy and adaptable palms we can grow here but, for me, it sits unhappily in the context of a dry garden. I would consider perfect partners for it to be plants like bamboo, schefflera, fatsia and other wet-climate Asian plants. Don't get me wrong, I absolutely adore them and, in the more jungle-orientated parts of my garden, I have over 30 planted out, as well as *Trachycarpus wagnerianus* and *Trachycarpus princeps*, but in my xeriscapes I am looking for something more reminiscent of hot, dry countries. They do tolerate my dry conditions but always look happier when receiving more rainfall.

Trithrinax

A small genus of fan palms from open, dry, rocky places in South America, distinguished by tough, stiff and often sharp-tipped leaves with sharp, spiny fibres clothing the trunks. For me there is a certain almost indefinable quality that makes these suitable for the dry garden. The spiny trunks help, of course, but seeing pictures of them growing in their semi-arid homelands it somehow makes sense, even though they are, at first glance, 'just' another fan palm. I grow two species, one of which, *Trithrinax acanthocoma*, has been in the garden since the late 1990s. It is a solitary palm with extremely stiff, mid-green leaves and a trunk clothed in a coarse webbing of dark brown fibre that ends in the most wicked spines. It grows steadily but slowly and has taken my worst three winters with only the slightest damage to older leaves. As it ages this palm develops more and more character, bringing its own identity to the fan palm family. I have tried to grow *Trithrinax campestris* three times. The first two attempts were imported trunked specimens but they both failed to establish, never re-rooting. It is a little sad to think of how many suffered the same fate, as there were quite a lot imported into the UK back in the early 2000s, although you rarely see one now. My final attempt has been in the garden for around ten years – a much smaller, seed-grown plant but one that had a complete rootball. This has been much more successful and is growing away happily, if slowly. This palm is a clumping species that enjoys a similarly vicious network of spines on the trunk as *Trithrinax acanthocoma* and, if anything, even stiffer, more dangerously sharp-tipped leaves of a glaucous grey-green. The overall effect is similar to a small but spiny *Chamaerops humilis* but with a harder, spinier edge to them that has added interest. (H4)

Trithrinax acanthocoma.

Trithrinax acanthocoma closer look at spiny trunk.

Trithrinax campestris at Tremenheere Sculpture Gardens.

Washingtonia

A genus of two species of large, fast-growing and, ultimately, extremely tall fan palms that are one of the most popular ornamental palms throughout the world, wherever the climate allows. *Washingtonia filifera* is native to the south-western USA and is a serious chunk of a plant with a hugely fat trunk that can reach 1.2m across at the base, 60m high at maturity and massive leaves that can reach 2m across. Leaf segments have conspicuous curly filaments along the margins and the petioles are armed with rather spiteful, sharp teeth. Of the two, this is said to be hardier to cold but less tolerant of moisture. *Washingtonia robusta* comes from north-west Mexico and differs in that the trunk is, at 90m high, ultimately taller but a lot thinner, the leaves smaller and not as noticeably filiferous. The base of the petiole, where it joins the trunk, generally shows a reddish-brown colour and the teeth are generally also darker. This species is said to be less hardy to cold but more moisture-tolerant and is the familiar skyduster palm lining the boulevards and avenues of Hollywood. In cultivation, hybrids between the two, known as *Washingtonia × filibusta*, abound and, in theory, these should combine the best attributes of each parent. It is often difficult to know, however, exactly which of the three you are buying, whatever the label states. The tell-tale characteristics are not always consistent, especially in younger plants, but with perseverance it is possible to track down seed-raised plants of known provenance. Examples of all three can be found in milder locations around the UK but temperatures below –6°C will usually damage the leaves and much below –10°C can kill the plants. They have been documented as growing back even if completely defoliated, however, so it is always prudent to wait a while if yours is not looking great come spring. I have one that I am confident is *Washingtonia robusta* that has been growing here for around eighteen years and which now has a 4.5m trunk. It lost all its leaves in my worst winter and took some damage in the early and late winters, but has always grown back quickly to make a completely new crown of leaves by the end of the following year. There is a giant specimen of *Washingtonia filifera* about 8km (5 miles) from me as the crow flies on Canvey Island that has been in a front garden there for nearly 25 years and is now 8m high. (H4/5?)

Massive *Washingtonia filifera* on Canvey Island.

Washingtonia robusta.

Xanthorrhoea

Over the years various grass tree species have been imported into the UK with very few happy endings. One of the biggest issues was getting them to re-root and I imagine most, if not all, of the imports of *Xanthorrhoea preissii*, *Xanthorrhoea johnsonii* and *Xanthorrhoea australis* from the late 1990s have died. The more recent imports, however, have been of *Xanthorrhoea glauca*, which seems far more adaptable to the UK climate and certainly easier to root. I planted one in the summer of 2010 and it has taken my early and late cold winters without damage. I do feel that it would need protection for temperatures much below –6°C. In habitat most grass trees experience flash fires, the shaggy petticoat of dried leaves quickly igniting and the flames soon moving on to the next plant. This gives them their characteristic scorched appearance and is something that can be tentatively recreated in the garden. I have recently done this to my *Xanthorrhoea glauca*, carefully snipping back the older leaves then, with a hose on stand-by, taking a blow torch to the stubble that neatly burned level to the trunk in the manner of a joss stick. The end result was pleasing, the newly blackened trunks making a strong contrast with the pale foliage. My friend Neil Armstrong, at Tremenheere Gardens near Penzance, has a number of magnificent *Xanthorrhoea glauca* – the finest grass trees in the country. He decided to do the same but, rather than take any particular care, simply lit a bonfire under each plant. The end result was the same, although, had that been me, my stress levels would have been off the scale. (H4)

Xanthorrhoea glauca at Tremenheere Sculpture Gardens.

A Close Encounter of the Xanthorrhoea Kind

In 1996 I was fortunate to acquire a beautiful multi-headed specimen of *Xanthorrhoea drummondii* – I used to help out at the nursery that imported them and had my pick of the new arrivals. As they were being unloaded from the shipping container, one particular *Xanthorrhoea drummondii* fell over in the container and out of its bag, landing squarely on top of me and leaving me trapped underneath for about half an hour until it was finally lifted off. I carried an interesting criss-cross scar on my back for many years from where the patterned trunk left its mark. During my brief incarceration I was able to closely examine the exposed healthy roots that this particular individual had grown and it seemed that Fate was telling me this was the one for me. It grew well in the garden for thirteen years but died during my worst winter. I still feel huge remorse about this as I am sure it was only for the want of a simple fleece blanket to stop the snow getting into the crowns.

Xanthorrhoea drummondii after a charring.

BROMELIADS

There are some 2,500 species of bromeliads in over 50 genera distributed largely through tropical Central and South America. A few are found in temperate regions of Argentina, Chile and Brazil, however, and it is here that we look for our garden plants. Typically, bromeliads grow on trees (epiphytes), on rocks (lithophytes) or in the ground (terrestrial): sometimes all three. Indeed, some tillandsia will grow on anything that does not move. The hardier bromeliads are usually terrestrial species that, to a greater or lesser extent, resemble outsized pineapple tops and are frequently armed with viciously hooked teeth along their leaf margins. Found naturally in arid places, often at high altitudes, in cultivation they need exceptionally sharp drainage and are perfect companions for the plants in this book.

Aechmea

One of horticulture's most closely guarded secrets is that there is a cold-hardy *Aechmea*: *Aechmea recurvata*. Hard, spiky, upright, urn-shaped rosettes, constricted a little at the neck, produce a magenta and cerise inflorescence, at which time the inner leaves, sometimes the whole plant, can flush anything from pillar-box red to a foxy russet. It will take –8°C in its stride, a lot lower if covered or protected, and will grow terrestrially or epiphytically. I met someone from east London who told me that he had grown it under a house brick on his garage roof

Aechmea recurvata.

Aechmea recurvata var. *ortgiesii*.

for the past 35 years, so these are tough plants! (H4)

There are also named varieties, var. *ortgiesii* and var. *benrathii*, selected cultivars, such as 'Big Mamma' and 'Big Red', and hybrids, such as *Aechmea* 'Covata' and *Aechmea* 'Sueños', although the latter two seem a bit more tender. All are extremely exciting. There are other species that show promise for milder regions such as *Aechmea kertesziae*, *Aechmea distichantha*, *Aechmea caudata* and *Aechmea gamosepala*, all apparently hardy to –5°C with overhead cover. I did, however, lose all these in my worst winter.

Billbergia

Billbergia nutans has been grown outside in the UK for some years. It is a small, delicate-looking plant best grown epiphytically, though it will grow happily in soil as long as it is freely draining. Upright, tubular rosettes of lax, bright green leaves, to 25cm long, hold a small tank of water. In late spring, nodding pink inflorescences charmingly sport pendant green and blue bells. Reliably hardy down to –8°C, maybe lower with overhead cover, I have grown a clump of this in the crotch of a yew tree for 25 years. It prefers to be kept out of the hottest, brightest sun, which can cause the leaves to turn yellow and even burn a little in dry periods. (H4)

Billbergia nutans.

I have recently been given a much chunkier, spinier form that, at first glance, does not resemble the usual type but the flowers are identical. Hopefully it will be as hardy, as it is a cracking plant.

Deuterocohnia

An honorary mention goes to *Deuterocohnia brevifolia*, which sits happily in a tight crevice in my vertical rock wall with a slight overhang that stops the weather from landing directly on it. I am not certain it would survive being grown out in the open but I am pleased to have it growing in its niche. This is a tiny, stiff-rosetted plant with short, hard, triangular leaves that, in nature, forms huge mounding colonies up to 2m across in the arid high Andes of Argentina and Bolivia. An individual rosette is just 2.5cm across and the flowers are curious luminous green, extended cones. Although I had grown this as a potted plant for many years, I first encountered it as a hardy plant when visiting a garden in the Netherlands. The owner had built his house with greatly extended eaves so he could have a range of cold-hardy but moisture-intolerant plants growing outside, including this little bromeliad. Winter temperatures in that area were usually in the –15°C to –20°C range. (H5 if dry)

I am also trying *Deuterocohnia lorentziana*, which, in looks, is almost exactly the same but nearly twice as large. Early days, but it does not seem quite as tough: exposed bits showed quite extensive damage after my early cold winter.

Deuterocohnia brevifolia.

Dyckia

There are many species, cultivars and hybrids of *Dyckia* available in the USA that, were they available over here, would doubtless prove to be hardy. Of those I have tested here, *Dyckia* 'Morris Hobbs' seems a good choice. It has 30cm rosettes of silvery leaves, turning burgundy with higher light intensity, and bright orange flowers on a simple spike in early summer. It has been hardy to –8°C in my garden, growing tightly wedged inside a rock crevice. (H4)

The same is true for *Dyckia velascana*, a beautiful silver plant from Argentina

Dyckia 'Morris Hobbs'.

Dyckia velascana.

Dyckia brevifolia.

distinguished by yellow flowers. The bronze-flushed cultivar *Dyckia* 'Cherry Coke' has survived my late and early cold winters, similarly growing in a rock crevice. I planted *Dyckia brevifolia* a few years ago in a border that was covered in winter and like that it came through –6°C in my late cold winter. Uncovered, however, it was killed by the same temperature in my early cold winter. Lastly, *Dyckia remotiflora* has been outside for many years, taking my worst three winters, but never really growing very much. I aim to move it to a different place to see if it will get going.

Fascicularia

This genus has been reduced from four species down to just one with two recognised subspecies. I use the word 'recognised' in a taxonomic sense, as there is enormous confusion over which plants are which in the gardening world, with many pundits failing to recognise or distinguish between the two subspecies. *Fascicularia bicolor* subsp. *canaliculata* (formerly *Fascicularia bicolor*) forms congested clumps made up of rosettes up to 120cm across, each comprising narrow, wiry leaves – greenish-grey above and silvery below. In cross-section they are quite thin and distinctly U-shaped. This one flowers freely, even in shade. *Fascicularia bicolor* subsp. *bicolor* (formerly *Fascicularia pitcairniifolia*) has much broader, more succulent leaves that are somewhat shorter – rosettes would be typically 80 cm – and leaves have a squat diamond cross-section. This subsp. *bicolor* is commonly seen all over the West Country and, in my experience, is shy of flowering. Both behave similarly when flowering – the inner circle of leaves flush bright, sealing-wax scarlet and out of a bulging, central, grey, pincushion-like bract peep scores of short, tubular, powder-blue flowers with bright yellow stamens. The inflorescence of subsp. *canaliculata* is broader and flatter, that of subsp. *bicolor* is narrower and raised, the floral tubes a bit longer. Each rosette dies after flowering but generally produces many offsets before doing so, thus enlarging the spreading clump. The biggest difference between the two subspecies is in hardiness – subsp. *canaliculata* will survive –12°C (H5) unscathed; subsp. *bicolor* is damaged below –6°C. Both are best sited in a sunny position but will grow happily in dry shade, as long as the soil is well drained. I find subsp. *canaliculata* especially useful as ground cover in dry shade as it is a plant my badgers tend to leave alone. They will both grow epiphytically if tied to the crotch of a tree and given a little sphagnum moss around the roots. I acquired a plant with the name *Fascicularia litoralis* but, other than as a synonym of *Ochagavia litoralis*, the only reference to this name I can find is via the website of the suppliers. I think it is a form of subsp. *bicolor*, but perhaps a slightly hardier clone that flowers more readily. Clumps can be divided to make more plants. I find it is best not to be too greedy and only divide a large plant into quarters. I have attempted several times to separate out individual rosettes and failed to root any

Fascicularia bicolor subsp. *canaliculata* flower detail.

Fascicularia bicolor subsp. *canaliculata*.

Fascicularia bicolor subsp. *bicolor* flower detail.

Fascicularia bicolor subsp. *bicolor* after –8°C.

Fascicularia bicolor subsp. *bicolor*.

Fascicularia bicolor subsp. *canaliculata* growing epiphytically in a Luton garden.

Leaf comparison between the two forms.

of them. If you do decide to lift and divide, wear long sleeves and thick gloves. Although not possessed of the hooked marginal teeth of puya leaves, those of fascicularia bear finely serrated margins that can cause quite a nasty raised red welt on exposed skin.

Greigia

Greigia sphacelata is a very interesting plant. It strongly resembles a larger, harder-leafed version of one of the South American eryngiums, such as *Eryngium agavifolium* until it flowers – strange, inconspicuous candy-pink things nestled within spiny pouches that are, themselves, hidden away in the leaf axils. These flowers are cute, in their own way, once the effort to seek them out is made. Most bromeliads flower from the centre of a rosette, which then dies, but this species continues growth on the same crown, which in time develops a short trunk. I have seen an ancient specimen of a closely related plant, which I believe to be *Greigia landbeckii*, at Logan Botanic Gardens, near Stranraer in Scotland, that was sporting a 1.5m trunk. *Greigia sphacelata* grows many offsets so that, eventually, the plant can take up an enormous amount of space: my oldest plant has ten crowns and spreads over an area

Greigia sphacelata.

Greigia sphacelata flower detail.

of 2.5m or more and reaches 1.5m high. The leaves, although hard, are fairly brittle and easy to crease along their length; also, the edges are armed with tiny but nasty little teeth that detach as soon as they get hooked into the skin. Thus plants are best grown away from main pathways and regular footfall. My own plant, which has been in the ground for twenty years now, grew much bigger than I anticipated and, of course, now encroaches over one of my main paths to the extent that it is hard to go past without picking up a spiny tooth or two. Short trousers are out of the question unless making a detour. There are big old specimens to be found in the Hillier Arboretum, Hampshire, in the gravel border opposite the house and also Queen Mary Gardens in Falmouth, Cornwall. I have read that in nature these are often to be found growing in trees as epiphytes, so I can only imagine they are nice sturdy trees, as these bromeliads are big heavy things once they get older. Hardy to –10°C or lower. (H4/5)

Ochagavia

A genus of four species, three of which are of interest and, of those, *Ochagavia andina* has been grown in the UK for the longest time. It can be found covering vast areas on the back terraces of the Abbey Gardens on Tresco, from where it has also escaped and naturalised on the sand dunes away from the confines of the garden. Historically this was believed to be *Ochagavia carnea* and this is the name under which it was distributed throughout the horticultural trade. Then, around eight years ago, plants raised from wild-origin Chilean seed bearing the name *Ochagavia carnea* became available, which were very evidently a different species. Checking more closely with the descriptions in published papers it became apparent that this new plant was actually the 'real' *Ochagavia carnea* and that the Tresco plant was the extremely rare *Ochagavia andina*, being conserved quite happily in the Abbey Gardens, unbeknown to the rest of the world. *Ochagavia andina* looks like a cross between a fascicularia and a puya, forming a mounding colony of fearsomely toothed rosettes that are, individually, around 90cm across. The leaf reverses are covered in patchy, silvery scales. In late summer an inflorescence resembling a pink baby's rattle adorned with cerise flowers is thrust out centrally from a rosette on a short stem. This species is hardy to –10°C or lower. (H4/5)

Ochagavia carnea has greener, flatter leaves that do not bear the silvery scales on the reverse. Rosettes are smaller but similarly produce many offsets to form a mounding colony. The large inflorescence, produced in May/June, is half as big again and nestled much closer to the centre of the rosette. Individual flowers are a similar bright pink, the same colour as the bracts. Hardiness has not been tested to the same extent but my plants have taken my late and early cold winters with no damage. (H4/5?)

Ochagavia andina.

Ochagavia andina flower detail.

Ochagavia carnea.

Ochagavia carnea flower detail.

Ochagavia litoralis.

Ochagavia litoralis flower detail.

The third species is *Ochagavia litoralis*, a smaller, neater plant all round with shorter, strongly recurving leaves, with a very dense silvery back. Older plants form a compact dome that is less scruffy than the other two. The inflorescence, appearing in August, is smaller and slightly furry, in texture like a baby's fleece blanket, with very pale pink flowers. Disappointingly this species is the least hardy, my plants showing damage (but recovering) at –6°C. (H4) If your climate allows this is the pick of the species for me. (H3/4)

Puya

Puyas must rank among the most viciously armed plants on the planet. Many have spiteful curved teeth along the leaf margins, hooked inwards near the centre, preventing easy or intact extraction of wayward hands – an adaptation thought to trap birds or rodents in a hostile native environment. The flowers of many species, however, are simply spectacular and worth the occasional blood sacrifice.

The most appropriate garden species is the relatively compact *Puya alpestris*. Individual rosettes reach 80cm in diameter and the plant will, in time, clump to 1.5m or more. It is a spectacularly flowering species producing a spiny 1.5m-tall, club-like inflorescence holding hundreds of waxy, metallic turquoise flowers – variably from almost black to much paler – studded with bright orange stamens. Hardy to –8°C if well sited, lower if covered. (H4/5)

Puya alpestris subsp. *zoellneri* (syn. *Puya berteroniana*, *Puya berteroana*) is equally hardy, similar in looks but half as big again in all dimensions. It is perhaps

Fabulous dark form of *Puya alpestris* at East Ruston Old Vicarage.

Puya alpestris subsp. *zoellneri* at East Ruston Old Vicarage.

Puya chilensis on St Michael's Mount.

Puya chilensis flower detail.

My *Puya* hybrid.

less inclined to flower. *Puya chilensis* is another monster with slightly straighter leaves and acid-yellow flowers, which is perhaps a touch more tender. (H4)

All three will branch and form multiple crowns – sometimes after flowering, often just routinely – making large colonies. Tempting though the prospect of those massive 2.5m inflorescences may be, these giants need serious consideration before planting as they take up a large amount of space and are exceptionally spiny. There is a common, as yet unnamed, hybrid between *Puya chilensis* and *Puya alpestris* subsp. *zoellneri*, often spontaneously appearing where both species are grown together, such as at the Ventnor Botanic Gardens on the Isle of Wight. I bought a plant of this, believing it to be *Puya berteroniana*, and it has been outside in my garden for 25 years, taking a minor hit in my worst winter but otherwise undamaged. It has flowered for me on five occasions, one time producing three inflorescences. From ground level to the top of the tallest spike was 3.5m and it was a spectacle that drew interested parties from far and wide. To give the plant a general tidy up prior to visits, I decided to trim the lower leaves and expose the snaking trunks and this has become an attractive feature of the plant. It is not a task to be undertaken lightly, it has to be said, and one that cost me a part of my left ear that ended up caught on a spine in a moment of carelessness.

Hybrid *Puya* flower detail.

I bought a large-trunked plant labelled *Puya coerulea* many years ago that may or may not be that taxon. It has been undamaged by my three worst winters and has flowered twice. It has the brightest silvery leaves of any plant I grow and, on bright moonlit nights, it absolutely shines. The flowers are an astonishing sapphire-purple with a metallic green sheen, held on a pinky-red stem. Marginal teeth on this species are not quite as evident as some but equally spiteful. (H4)

Puya spathacea is almost certainly the hardiest of the genus, taking my three worst winters in its stride, but in other, colder, gardens taking below –10°C without damage. The leaves are wide and relatively short with wiry marginal prickles. The flowers are deep turquoise – almost black – peeking out modestly from covering bracts and held on colourful stems that can vary from pale pink to radish-red. (H5)

Puya coerulea.

Puya coerulea flower detail.

Puya spathacea.

Puya spathacea flower detail.

Puya harmsii.

Puya harmsii is similar but with narrower, more silvery leaves and a larger, more branched inflorescence. Flowers are even darker inky-purple, almost black, but with a hint of metallic green. Probably not as hardy. (H4)

Puya assurgens came through my worst and late cold winters just fine but the leaves were spotted by the early cold winter. It is, overall, similar to *Puya spathacea* but with softer leaves and smaller lateral branches on the inflorescence. (H4)

I have only been growing *Puya dyckioides* for a few years but it impresses me, taking my late and early cold winters without issue. The flowers are a kind of metallic jade-green, just about poking out from the colourful pinky-red bracts on arching stems. This species takes more shade than the others. (H4)

Puya mirabilis is fairly commonly available but, for me, is nearly always disappointing. The flowers appear so late in the year that they rarely make it before the first frosts and, when they do, are usually eaten through by snails before fully opening. But, when the flowers do manage to open, they remind me of the reason I persist with growing it – they are very large, three-petalled, pale green, lily-shaped bells that are produced in a nodding cluster and simply gorgeous. (H4)

I have one *Puya* that has yet to flower and confirm its identity. It has come through my three worst winters completely unscathed and has gradually grown into a large and impressive clump of wide, silvery leaves armed with reduced, but still nasty, spines spaced widely along the margins. I suspect it is *Puya castellanosii*.

Puya assurgens.

Puya dyckioides.

Puya dyckioides flower detail.

Tillandsia

A final paragraph goes to *Tillandsia*. These are the 'air plants' seen glued to seashells in garden centres all over the country. Surprisingly, perhaps, a few grow well outside. *Tillandsia aeranthos* is a robust plant that forms a pendant, creeping and branching stem of recurved, silvery, minutely hairy leaves that will eventually grow into quite dense balls. It has strong-coloured purple flowers that emerge from pink-red bracts. *Tillandsia bergeri* is a very similar but smaller plant with pale mauve flowers from paler pink bracts. *Tillandsia albertiana* looks rather different, resembling a bundle of dark green pine needles tied together producing unexpectedly showy scarlet red flowers. These really could not be easier to keep – just nail, tie or glue them to pretty well anything to stop them blowing away, give them a spray with water every couple of days in summer and a fortnightly spray with a liquid feed. My friend Tony King grows them to great effect in the leaf base pockets of palms where they establish and grow quite naturally, enjoying the protection of the evergreen canopy of palm leaves. All three are hardy to –6°C or lower. (H4)

Assorted tillandsias.

Tillandsia albertiana.

If permanent planting sounds risky or inappropriate, practically any bromeliad species can be pot-grown and placed outside for spring through to autumn. The epiphytes are particularly accommodating and can be grown in hanging baskets, attached to old tree fern trunks, driftwood or anything that fires your imagination. They can then be moved under cover and kept just frost-free for winter. Many *Neoregelia* and *Aechmea* species, which include some of the showiest foliage colours of all bromeliads, are very robust and easy plants to keep in this way, requiring little by way of care but rewarding with sumptuously exotic colours.

TREES AND SHRUBS

Acacia

A large genus of trees and shrubs found in Australasia, Africa and North America, these are one of the signature trees of arid scrubland. Certainly I have struggled my way through sweet thorn acacia in Mexico more times than I care to remember and have the rips in my clothing to show for it. The leaves are interesting. Many species have feathery bipinnate leaves; others have flattened, willow-like leaves that are actually modified leaf petioles known as phyllodes. The flowers are fabulous clusters of yellow pompoms, often scented, and in full bloom during late winter to spring at a time when colour is most welcome. As a rule, the buds form during late summer and sit there waiting for the right weather to come along, so it is not unusual, in extremely mild winters, for an acacia to be in flower on Christmas Day. Several species are hardy enough to grow throughout much of the UK and a few of those have the perfect 'wispy desert tree' look that sits so comfortably in the xeric garden.

My absolute favourite is *Acacia boormanii* from south-east Australia, which grows into a weeping shrub, eventually around 5m tall, that will fit into any sized garden. The evergreen phyllodes are 8cm long, grey-green (some forms are a striking blue-grey) and very narrow: the whole plant is covered in masses of lemon-yellow flowers in late winter. A well-drained, even poor, stony soil in full sun will suit it best. This really has it all: drought-tolerance, hardiness to cold, imperviousness to heat, colourful flowers and takes the gold medal for a wispy desert tree for the UK climate. (H4/5)

Acacia *boormanii* flower detail.

Acacia dealbata subsp. *subalpina* in flower.

Acacia boormanii.

Acacia dealbata subsp. *subalpina* leaf detail.

Acacia dealbata is one of the hardiest but a huge 30m monster. There is a subspecies *subalpina*, however, that tops out around 5m high, much more suitable for domestic gardens, which is also a high-altitude form, so a few degrees hardier. Leaves are glaucous blue-grey and bipinnate. The flowers, borne in late winter, are bright yellow. This is, for me, the finest of the feather-leafed acacias. (H5)

Acacia obliquinervia is one of the willow-leafed acacias. Its 8cm-long, glaucous grey phyllodes are much broader than those of *Acacia boormanii*, and are, at first glance, strongly reminiscent of eucalyptus leaves but the clusters of golden pompom flowers soon put paid

Acacia obliquinervia.

Acacia pravissima.

to that idea. This is an absolute gem of a plant, known to survive down to –15°C, yet it is unaccountably rare in cultivation. Mine has been in the garden for twenty years, yet is still only 3m high and broad. (H5)

Acacia pravissima is a small tree to around 6m tall with a weeping habit and strange triangular-shaped, dark-green phyllodes. Masses of bright yellow, scented flowers are produced in, usually, March. This is one of the hardiest of the acacias, taking my worst and late cold winters in its stride, only to suddenly die back for no apparent reason. I have planted a replacement. (H5)

Finally, I mention *Acacia verticillata*. This is another unusual acacia, growing to maybe 8m tall, in that the phyllodes are reduced to long, sharp needles, bundled together in whorls along the stems. The pale yellow flowers are held in tight, cylindrical, bottlebrush clusters giving a different appearance to most. It is, sadly, not as hardy as the other species above: I have grown it twice, losing my first plant in my worst winter, and the second plant in my early cold winter. For mild regions, especially near the coast, this is definitely worth trying. (H3)

Albizia

A genus of over 160 species, related to *Acacia*, many of which come from dry, arid climates. One has become a fairly commonly grown ornamental plant in the UK. *Albizia julibrissin* has much to commend it, growing a wide-spreading, umbrella canopy, at maturity 10m high and potentially the same wide, with graceful, almost arching, deciduous bipinnate leaves around 30cm long. All through summer the tops of the branches are covered in erect, pink, sweetly scented, shaving-brush flowers. Flower colour and size can vary: there are selected cultivars that make better garden plants than the 'plain' species. 'Rosea', sometimes seen as forma *rosea*, has flowers that are, as the name suggests, a stronger pink. 'Ombrella' has darker flowers again, making a more compact tree at 5m high. It lends an extremely exotic air to the xeric garden, the delicate leaves contrasting wonderfully with the larger, spiky plants and palms. It is by no means perfect, being extremely late to leaf out, sometimes as late as July. The flowers do not die gracefully and fall off neatly but hang around looking brown and scruffy just that little bit too long. A happy tree in full flower, however, is a sight to behold, especially on still evenings when the sweet fragrance can pervade the whole garden. I grew this for over twenty years, making a wonderfully elegant 8m tree until it finally succumbed to coral spot fungus. I understand these grow better in the east and south of the UK, where they are hardy to –10°C or lower. (H4/5)

Albizia julibrissin.

Arbutus

A wonderfully ornamental genus, comprising a dozen species from North America, the Mediterranean, Western Europe and the Canary Islands, *Arbutus* is a small evergreen tree with attractive peeling bark, often revealing a smooth new layer beneath. In the same family as heather, these trees have similar bell-shaped flowers that develop into edible fruits, reminiscent of strawberries, that

Arbutus 'Marina'.

The wonderfully tactile bark of *Arbutus* × *andrachnoides* at Glasnevin Gardens.

hang in prominent clusters along the branches. *Arbutus unedo* is found wild in many Mediterranean countries and also, counter-intuitively, a couple of locations in Ireland. They make small trees with beautifully fissured bark and, when established, are incredibly drought-tolerant. *Arbutus andrachne* is the eastern species, from Greece and the Middle East, and has peeling bark that reveals the smoothest, most tactile russet-coloured trunk. Where these two grow together they form a natural hybrid *Arbutus* × *andrachnoides.* Another hybrid, of uncertain parentage but postulated to be *Arbutus*× *andrachnoides* × *Arbutus canariensis*, was discovered in a Californian nursery in the 1920s. Grown locally for 60 years, it was finally distributed under the cultivar name *Arbutus* 'Marina' in 1984. This, for me, is the pick of the genus and the one I have had growing here for twenty years. Compared to the others, *Arbutus* 'Marina' has peeling bark, bigger leaves, more prominent flowers and larger fruits. It has never been troubled by cold, heat or drought. Any of them are worth growing. They always have immense character, ageing very quickly and developing craggy, gnarly branches within just a few years. (H4/5)

Brachychiton

This consists of a handful of ornamental trees from Australia, a couple of which show some hardiness. *Brachychiton populneus* is the hardiest species, an evergreen tree forming a robust trunk with dense branches of narrowly pointed, elliptical leaves and clusters of nodding, red flowers. Grown from seed, plants are variable – juvenile foliage often being three- or five-lobed. This was slightly damaged by my early cold winter but grew back vigorously. *Brachychiton rupestris* is more interesting, forming a bottle-shaped trunk with attractively deeply cut, evergreen leaves but, sadly, less hardy. Defoliated by my early cold winter, it has sprouted from low on the trunk. (H3/4)

Unusually cut-leaf form of *Brachychiton populneus*.

Casuarina

Distributed around Oceana, *Casuarina* is a small genus of evergreen trees, generally called she-oaks, that have whipcord foliage, the leaves being reduced to small scales clasping the twigs, giving an over-

Casuarina cunninghamiana.

Male flower detail on *Casuarina cunninghamiana*.

all effect of a conifer of some sort. A few species are grown as ornamental trees in arid climates around the world, one of which, *Casuarina cunninghamiana*, has proved to be reliably hardy for me for the past 25 years and is now 8m tall. In that time it has taken cold, wet, drought and heat with absolutely no ill effects. When it was just a small plant, someone (not me! – a neighbour) trod on it, snapping the stem almost halfway through. I set it back up straight and then, in the absence of anything more suitable, held the break together with gaffer tape, a repair that miraculously mended the damage and has endured to this day. In 2018, Storm Eleanor caused a few problems in my garden, not the least of which was a 12m-long piece of ash tree that was blown unceremoniously on top of the *Casuarina*, bending its trunk to a right angle but saved from complete destruction by a large *Pittosporum* 'Irene Patterson'. A few hours' work with a pruning saw released the trunk, which then sprang back up as though nothing had happened. Clearly these are tough plants. Sexes are on separate plants; mine is a boy and releases huge clouds of pollen over the course of a week or so, covering everything nearby in a dusty layer. (H4)

Ancient carob tree in Portugal.

Ceratonia

Only the species, *Ceratonia siliqua*, concerns us. This is the carob tree commonly seen all over the Mediterranean, where they are native, especially in Portugal. A robust evergreen with leathery pinnate leaves and a rugged, textured bark, these impressive trees attain a great age in their habitat, with trunks often rotting in the middle and carrying on as twins for another few hundred years. Flowers are borne directly on the older wood and, on female trees, lead to huge, dark brown pods filled with reddish-brown seeds, which, when dried, roasted and ground, are the source of the ersatz cocoa powder, carob. The received wisdom is that they are not hardy; however, many years ago I heard of (but have yet to find) a carob tree of some decades' vintage, growing in a garden in Southend a few miles from where I live. Encouraged by this I bought a reasonably large plant and it has been growing happily for fifteen years, taking my worst three winters without any issue. How much lower it will take I do not know but, for sure, –8°C and a fortnight below freezing were OK. My plant is actually in a location that is too shady but, at the time, I did not have anywhere appropriate. Rather than attempt to move this one, I have just planted a second tree in a sunnier spot and look forward to seeing how that establishes. This is a plant definitely worth trying in a hotspot if you can spare the space. (H3)

Ceratonia siliqua.

Colletia

A small genus of thorny shrubs from South America with spiny stems and fragrant, white flowers. *Colletia hystrix* (syn. *Colletia armata*) is sometimes called the barbed wire bush – a name that fits the plant perfectly, as it makes a tangled

Colletia hystrix at the World Garden, Lullingstone.

Colletia hystrix pink form at the World Garden.

Colletia paradoxa at RHS Hyde Hall.

shrub, to 3m, of spitefully thorned stems with insignificant deciduous leaves. Small, tubular flowers borne in autumn are pervasively almond-scented and the last thing you would expect to find on such a scary-looking plant. (H4)

Colletia paradoxa (syn. *Colletia cruciata*) is the most formidable-looking plant, with broad triangular, spine-tipped, flattened stems, called cladodes, arranged in alternate pairs along the branches. Leaves are insignificant and deciduous. Flowers are similar to those of *Colletia hystrix* and it will reach 4m high in time. (H4)

These are just such odd-looking plants that, if space allows, are must-have plants, as they fit in with the other spiky plants so well.

Colletia hystrix flower detail.

Eucalyptus

There are hundreds of species of gum tree, all from Australasia; quite a few of them are hardy to grow in the UK, and some of them are huge monsters. I am going to mention just two of the smaller ones. *Eucalyptus pauciflora* subsp. *debeuzevillei* is for me the stand-out eucalyptus for the UK. It stays relatively small, topping out at around 10m. The bark is a wonderful python-skin pattern of white, cream, grey and brown. It has evergreen leaves that are long, broad and leathery; new shoots are crimson with a glaucous bloom changing to glaucous, silvery-blue. The whole plant is suffused with the typical delicious scent of eucalyptus oils when crushed or cut. White flowers are held in long clusters in late spring. If it outgrows the space allowed, it can be coppiced, as it will regenerate from a lignotuber near the base, an adaptation some gum trees have to allow regeneration after wildfires. It is drought-tolerant and hardy to at least –18°C, being one of

Eucalyptus pauciflora subsp. *debeuzevillei* python-skin trunk pattern.

Eucalyptus pauciflora subsp. *debeuzevillei* flowering.

Eucalyptus nicholii at RHS Hyde Hall.

Eucalyptus nicholii trunk detail.

the snow gums from as high an altitude as Australia gets. Can we possibly ask any more of a tree for our gardens? (H6)

There is a mature specimen of *Eucalyptus nicholii* at the RHS Garden at Hyde Hall that has been growing there for 50 or more years, taking all the heat, drought and freezing temperatures at the garden in that time. Known as the willow peppermint gum it has immense character and, even at that age, is an acceptably compact 12m or so tall. It does not have the eye-catching bark of the snow gums, but rough, ridged, reddish-brown, fibrous and peeling strips packed with character. The sage-green leaves, borne on red stems, are long and narrow, and when crushed have a mix of eucalyptus and peppermint scent. Clusters of white flowers are produced in late summer. The willow peppermint gum seems to have slipped under the radar somewhat, which is a great shame. (H5)

Euphorbia

Succulent spurges appeared earlier in the book but there are some absolutely magnificent shrubby or sub-shrubby species to join them. Three form a natural group: *Euphorbia mellifera*, from Madeira and the Canary Islands, and *Euphorbia stygiana* from the Azores, both grow in the

Euphorbia mellifera.

damp laurisilva forests present on the wetter parts of the islands, so it is rather surprising that they have pretty good drought-tolerance. There is a hybrid between the two, *Euphorbia × pasteurii*, with a couple of named cultivars that are, for my money, better garden plants than either parent. During a recent extended drought and heatwave my *Euphorbia mellifera*, a 25-year veteran, was completely defoliated – the only time that has ever happened. It regrew. The *Euphorbia stygiana* and *Euphorbia × pasteurii* fared better, albeit they were unimpressed by the weather. (H3/4)

Euphorbia characias subsp. *wulfenii*, often called the Mediterranean spurge, is a well-known, widely grown plant and with good reason. When the idea of 'gravel gardens' became fashionable these were suddenly ubiquitous and, still today, are unrivalled for architectural presence, bright spring colour and drought-tolerance. In fact, other than the irritant sap, which is there with any spurge, there is little to say against them. Various cultivars have been selected but the plain subsp. *wulfenii* is hard to beat in my opinion. The fat chartreuse-yellow flower heads hovering above the grey whorls of leaves is just such a winning combination. All this, and it is hardy to –15°C. (H5)

A hybrid between this and *Euphorbia amygdaloides* exists called *Euphorbia × martini*, which, for some unaccountable reason, I have only been growing for the past couple of years. It has quickly impressed with the maroon-tinted new growth, compact size and colourful heads of flowers – not as dense as *Euphorbia characias* subsp. *wulfenii* but perhaps more colourful, as the lime-green flowers have maroon eyes. A baptism of fire; I planted a group of five in a south-west-facing bed of rubble and mortar, and their first summer was the hottest and driest on record. They did not miss a beat and looked just as fresh by the time the first rains came as they did at the start. An added bonus is that flower heads turn a strong russet and stay in good shape through winter. (H5)

Another natural group are the sub-shrubby, sprawling species. *Euphorbia myrsinites* is perhaps the most familiar of these with its trailing stems carrying glaucous blue whorls of broad but short triangular leaves and acid-yellow flowers in early spring. (H5)

Less often seen, *Euphorbia rigida* is similar but more upright, flowering earlier in February/March. For me this species has more presence, as the longer triangular leaves are more rigid and densely packed, and the fading flowers and bracts have reddish notes, adding to its charm. I have this growing out of a vertical rock wall and it initially arches downwards then turns up at the tips lending a curious form to the whole plant. (H5)

Euphorbia veneris is very rarely seen; it comes from the Troodos Mountains in Cyprus and is somewhere halfway between the two previous species in looks, with narrower leaflets and a reddish tint to the bracts. (H4/5)

The last of this little group is the diminutive little creeper, *Euphorbia capitulata*,

Euphorbia × martinii flower detail.

Euphorbia × pasteurii.

Euphorbia characias subsp. *wulfenii*.

Euphorbia × martinii autumn colour from seed head.

Euphorbia myrsinites at RHS Hyde Hall.

Euphorbia rigida at RHS Hyde Hall.

Euphorbia veneris.

found growing in rocky limestone scree above the treeline in the Balkan mountains. Lax stems clothed with whorls of blue-grey, obovate leaves rarely reach 10cm high. In late spring the stem tips proudly display the typical spurge-like, yellow flowers, fading to russet. I lost the label many years ago and have only very recently rediscovered its ID but my plant has anonymously and slowly spread to a dense patch 30cm across. (H6)

Euphorbia capitulata.

I am finishing off spurges with two odd plants. *Euphorbia dendroides* comes from coastal Mediterranean locations, the Balearic Islands and east to North Africa, in time growing into domed woody-stemmed bushes up to 2m in height. Most references I have seen say this plant is extremely tender to cold but my experience is quite different. I collected some seed in Rapallo, Italy, in 2013 and planted one of the resultant seedlings in my south-facing rock wall where, to my surprise, it has been growing happily ever since, even self-seeding in the way that spurges do. It is summer-dormant, often completely shedding its leaves, only to seemingly miraculously resurrect after the first rains of autumn. Flowers appear in early spring. This plant has taken –6°C with a week of temperatures below freezing, coupled with a covering of wet snow, enduring both my late and early cold winters without any damage whatsoever, even as small self-sown seedlings. This has a certain something – a look that says 'hot and sunny' in a way that most spurges do not manage, so if you have somewhere hot and sunny with perfect drainage then it is well worth a try. (H3/4?)

Last mention goes to *Euphorbia margalidiana*. This is incredibly rare, found

Euphorbia dendroides flowering.

Euphorbia dendroides in Mallorca.

Euphorbia margalidiana flowering.

only on two chunks of rock off the coast of Ibiza called Ses Margalides. It strongly resembles *Euphorbia balsamifera*, one of the coastal Canary Island spurges, but it is not quite as woody, large or dense. It is similarly summer-dormant, flowering in spring, with leaves of similar shape and colour. It is, in my opinion, simply the most exotic of all the shrubby euphorbias I have grown. My plant grew happily for a dozen years or more, taking my worst winter in full leaf without a mark. I decided to widen the path next to where the plant was growing and moved it, an indignity from which it

Euphorbia margalidiana.

never recovered. With hindsight I should have just moved the path somewhere else. All was not lost, however, as I had previously passed on some cuttings to a friend who managed to root a couple and still had one remaining, if sickly, plant left. I planted it in a hot, sunny place with its roots in rubble but, unwittingly, on a route used by foxes. I was distraught to find it in bits one morning. Looking on the internet there are suppliers of *Euphorbia margalidiana* but the pictures of their stock show them to be something else entirely, the imposter having apparently been awarded the RHS Award of Garden Merit. Against all hope, however, I found a new source and now have a replacement planted in a hot sunny place with its roots in rubble and crumbled mortar, safely tucked away from any paths – human or vulpine. (H5)

Genista

A large genus of trees and shrubs in the pea family, many of which are drought-tolerant. Of these, *Genista aetnensis* is the stand-out species for the wispy desert-tree look. Weeping, almost leafless branches are covered in fragrant, yellow flowers all summer long. With age they develop immense character and there are wonderful old specimens to be seen in Beth Chatto's Gravel Garden in north-east Essex and also a magnificent solitary street tree on Marine Parade in Leigh-on-Sea, just a few miles east from me. For no reason I can remember, I have never got around to growing one here. I still intend to, although as they look their best once attaining twenty years or more, I think I'd better get a move on. Hardy to –15°C. (H5/6)

Genista aetnensis in Leigh-on-Sea.

Laurus

Laurels are a small genus of evergreen trees with aromatic foliage. *Laurus nobilis* is the familiar sweet bay or bay laurel, from the western Mediterranean and Middle East, the leaves of which have been used for centuries as a culinary herb. Some sources suggest these are somewhat tender but there are 10m mature plants dotted around south-east Essex and beyond that would contest this. I have a slightly rarer type, *Laurus nobilis* 'Angustifolia' (syn. *Laurus nobilis* forma *angustifolia*), which is now an impressive 6m columnar tree. This has much narrower leaves than the type and manages to maintain a certain elegance without any need to be shaped. When used in cooking I find the flavour imparted by the leaves to be sweeter and more fragrant than the type. (H4)

Laurus nobilis 'Angustifolia'.

Laurus nobilis 'Angustifolia', a closer look.

Leonotis

A tropical genus of shrubby plants in the mint family of which I am growing two species, both from Africa. With me, *Leonotis leonurus* remains partially evergreen but is best cut back in spring. Colder winters do this job for me but it generally returns. Square-section stems are clothed in aromatic, oblong, coarse-textured leaves. From late summer until the first frosts, strange whorls of burnt-orange flowers are produced from prickly balls along the upper stems. As an evergreen shrub in mild climates this can reach 2.5m high and across but as a returning perennial mine makes maybe 1.8m in a season. New to me is *Leonotis nepetifolia*, which has a similar arrangement of flowers in a similar colour but with diamond-shaped leaves, apparently growing to 3m in milder climates. I have yet to overwinter this one but if it proves to be as tough as its cousin, I think I

Leonotis leonurus.

Leonotis nepetifolia.

prefer it. In cold regions both of these species can be grown, and will flower, as annuals. For tall splashes of colour these are magnificent and easy-care plants, needing no irrigation or staking. (H3/4)

Mahonia

A genus containing some of my favourite evergreen shrubs for the exotic garden, there are a few species from the USA and Mexico that are particularly appropriate for the xeric side. Two species, *Mahonia haematocarpa* and *Mahonia trifoliolata* are rather similar, with very tough, leathery, extremely prickly evergreen, blue-green leaves densely clothing the stems and scented yellow flowers followed by red fruits. Arguably the most attractive is *Mahonia fremontii*, which has wider, not quite as thorny-edged, leaves that emerge red and eventually age to the most gorgeous powder-blue. Fruits are purple. All species must be given full sun and perfect drainage. They have a tendency to get a bit straggly but can happily take light pruning to keep them tidy. Certainly the neat plants I have seen in Mexico have had this done for them by goats, so it is unusual to see these shrubs reach their full size of around 3m. I grew *Mahonia haematocarpa* for several years, including through my worst winter, but it started to crowd out its neighbours. Rather than keep it trimmed, I decided to move it somewhere with more elbow

Mahonia haematocarpa.

Mahonia fremontii.

room and, of course, it died. I have recently planted *Mahonia fremontii* in a sunny spot in rubble and with ample space. (H4/5)

Melia

Melia azedarach is a familiar street tree for holiday destinations in hot countries, being drought- and heat-tolerant and casting light shade. Elegant, bipinnate leaves with ovate, toothy-margined leaflets and huge clusters of lilac flowers that, in hot summers, are followed by yellowish-orange berries. It is not widely grown but there is a huge specimen growing near the Chelsea Physic Garden in London. It would make a very interesting and unusual tree for those with hotspots. I have tried it twice – the first time it was slowly but inexorably ringbarked by snails that seemed to find the new growth irresistible. The second time my newly planted sapling had to come to terms with my early cold winter and died. There is one at the World Garden at Lullingstone that is still alive after −17°C, so I will keep trying! (H3).

Melia azedarach in Barcelona.

Melia azedarach in the World Garden, Lullingstone.

Nandina

Nandina domestica, native to eastern Asia, is the only species in this monotypic genus. An exceptionally useful evergreen shrub, related to *Mahonia*, with evergreen bipinnate or tripinnate leaves forming a domed canopy to 3m or more, providing a wonderful delicate texture that is unlike most drought-tolerant evergreens. Suckers appear from ground level but never enough to be a nuisance. New growth is copper-red and foliage can turn red over winter. There are sprays of creamy white flowers that, in warm sites, develop into eye-catching red berries. I consider it to be one of the most accommodating plants, growing in sun or shade, moist soil or dry, heat or cold. I have seen it planted in Arizona alongside dasylirion and yucca, and a friend reports seeing it in tropical Burma. There are many compact and colourful cultivars appearing on the market but, for me, nothing beats the plain species and I have many plants of this growing in various situations in my garden. There is, however, a catch. All parts of the plant are poisonous, especially the berries. Livestock and pets are at risk – cats do not eat berries but some dogs will eat anything. If yours does, do not plant one. (H5)

Nandina domestica.

Nerium

One of the signature plants of the Mediterranean, *Nerium oleander* has to get a mention. Often considered to be rather tender, the good news is that different forms and cultivars have a range of hardiness. There is a form collected from the Atlas Mountains in Morocco that is said to take down to −18°C; the cultivar 'Villa Romaine' is reputed to take down to −15°C. This is the one I grow and it has been fine during my late and early cold winters. Both these have single, pink flowers opening from red buds. I planted a fine pair of this cultivar at my sister's house, a few miles east of here, either side of the porch on the south-facing house wall, which have grown and flowered spectacularly well. Just be aware that oleander is one of the most poisonous plants you can have in the garden – all parts are extremely toxic. (H5)

Nicotiana

A slightly tender, shrubby member of the genus, *Nicotiana glauca* loosely resembles a small eucalyptus with leathery, glaucous-grey, elliptical leaves but bears, on the tips of the branches, sprays of yellow, tubular flowers. Even if not

Nicotiana glauca.

damaged by cold, established plants are best cut back in spring, making over 2m of bushy, fresh growth annually from the woody base. Not a long-lived plant but easily raised from seeds, which are produced in vast quantities once the flowers are spent. (H3/4)

Olea

Another 'must-have' plant from the Mediterranean, *Olea europaea* needs no description. Once thought to be tender, we now know these will take −15°C with little or no damage. My advice is to always buy the biggest and gnarliest trunk you can afford, rather than be swayed by a full canopy. The leaves and branches come and go and can be pruned however you want, but the trunk is always there. Plant it in the stoniest, driest, hottest part of the garden, water it in just the once, then forget about it until the crown needs shaping. I tend to prune mine every spring, cutting the branches back by two-thirds. If I miss a year, the canopy can get a bit sprawly for the space allotted to it, so in the second or third year I will cut the branches back hard to leafless stubs and let it start again. (H5)

Pistacia

The familiar mastic bush from the Mediterranean, *Pistacia lentiscus* is another plant that I have found to be much hardier than received wisdom would suggest. I have grown it here for over twenty years and it has taken my worst winters without damage. I am sure they would be slightly damaged at −10°C but equally sure they would recover. I have several, which I grow in both full sun and shade: they seem happy with either but retain a more compact shape with better light. The distinctly

Pistacia lentiscus.

Pistacia lentiscus starting to flower.

lobed, pinnate leaves, reddish when young, add a great contrasting texture and the plants I have in the sun have set the characteristic red berries before now. (H4)

I have yet to grow *Pistacia terebinthus* but aim to do so as soon as I track down plants for sale. It is similar to *Pistacia lentiscus* but larger and deciduous: the flower clusters are larger and more open, the berries more conspicuous. It should also be a bit hardier.

Ancient olives in Mallorca.

Pittosporum

There are scores of *Pittosporum* species and cultivars that make good, solid, if perhaps a little dull, garden plants but there are a couple that, for me, stand out as suitable for the exotic dry garden.

Pittosporum eugenioides

This is one of my favourite evergreen trees as it somehow seems to sit happily with any style of garden. The long, glossy, apple-green leaves with contrasting pale mid-ribs undulate beautifully along the margins, giving it a classy and exotic appearance. Dense clusters of little, highly scented, yellow, tubular flowers are produced in April. It is most usually seen in the variegated form, which I think is a shame as, in my opinion, the variega-

Pittosporum eugenioides.

tion detracts from the elegance of the foliage. I have been growing this for 25 years and it has never been affected by heat, cold, drought or wet. (H4/5)

Pittosporum tobira

This is much more commonly seen, especially in the Mediterranean region where it is an extremely popular ornamental. It is often seen tortured in tiny terracotta pots but it is at its best when given its head and allowed to grow into a small tree with a spreading canopy. Although mostly grown in full sun, I find it is happy in dry shade, those leathery, glossy green leaves reflecting the light beautifully. If this induces legginess, plants can be cut back, even into old wood, and will quickly regenerate. Strongly scented, creamy white flowers in summer are a bonus. This is hardy to around –10°C. (H4/5)

Pittosporum tobira.

Pseudopanax

A genus of incredibly useful Antipodean evergreens for exotic gardens that mostly suit the jungle style. A couple of species, the lancewoods, are so peculiarly individual that they look equally eye-catching in either a jungle or xeric setting; fortunately they are both drought-tolerant.

Pseudopanax crassifolius is a fascinating tree for all sorts of reasons. For the first fifteen or so years it grows as a single, or sparsely branched, vertical stem clothed with long, narrow, downward-pointing leaves that can reach anything up to 90cm and are edged intermittently with sharp, flat teeth. These leaves are very tough and leathery, and usually dark green with a prominent reddish-brown mid-rib. In their late teens to early twenties, the plants change appearance: the stem branches, the leaves get increasingly shorter and are held horizontally and then finally upwards, the whole plant taking on a more regular 'lollipop' tree appearance.

This strange behaviour is believed to be an adaptation to avoid being browsed by a large flightless bird, the giant moa, that used to live alongside the lancewood trees. It was postulated that once out of beak reach, the tree could safely grow normally. A difficult theory to prove as the giant moa has been extinct for some centuries. A very closely related plant, *Pseudopanax chathamica*, evolved on Chatham Island where there were no giant moa and it does not go through the changes. All of that aside, *Pseudopanax crassifolius* is an extraordinary looking plant that is both spiky and drought-tolerant, so finds a place here. I have two planted together that are

Pseudopanax crassifolius adult form.

Pseudopanax ferox in Eastbourne starting to change.

Pseudopanax ferox juvenile foliage.

Pseudopanax chathamicus at RHS Hyde Hall, no foliar Tomfoolery.

currently going through 'the change' and I am considering pollarding them as I understand this will make the tree revert to the more curious juvenile form. Then again I do have a track record for killing things by messing them around, so maybe I will leave them be.

Pseudopanax ferox follows a similar growth cycle but has even fiercer leaves, with more numerous marginal rip-saw teeth, that are shorter and dark brown. It grows much more slowly and, if anything, is even more of a curiosity.

Both are H4/5.

Ruscus

A small genus of plants related to the woody lilies described earlier. They are small shrubs with branched stems that spread from creeping underground rhizomes, eventually making large colonies. They have what look like tough, leathery, evergreen leaves that are actually flattened stems called cladodes. The centre of each cladode bears a tiny vestigial proper leaf at the reduced 'branch' and it is also where the flowers and berries are borne. Some of them are amongst the very finest colonisers of dry shade that it is possible to grow.

Ruscus aculeatus is native to Eurasia, its range extending as far west as the UK. It is representative of what the group looks like, with branched stems to 80cm long, carrying open whirls of cladodes to 6cm long, and armed with a pointed tip that is sharp enough to draw blood. Although uncommon, it can be found growing wild in my region, there being a couple of huge patches of it on the route of one of my regular walks. Interestingly, both patches are male, but one corner of one patch bears female stems and regularly carries berries, a rare but not unique occurrence. These are spiny, prickly, drought-tolerant, evergreen, indestructible plants that are totally indispensable in the garden. (H6)

I love them: all the various forms and species. Yet I realise from the large number of my friends who do not see the value in them that I am in a minority. *Ruscus hypoglossum* is much the same size and configuration but an altogether more user-friendly plant, the 8cm long cladodes lacking the sharp tips of those of *Ruscus aculeatus*. (H5)

Ruscus colchicus is similar but much larger, the cladodes reaching 13cm in length. (H5)

The jewel in the *Ruscus* crown is *Ruscus hyrcanus*, which is one of those plants that is simply impossible to obtain by normal means. Its range in the wild is those countries from Georgia to Iran via Azerbaijan that are bordered by the Black Sea to the west and the Caspian Sea to the east – places that are unsafe and off the agenda for travelling at the moment. I understand that there is one single clone in cultivation that was collected in the mountains of northern Iran by a US serviceman when he was stationed there in the 1970s that has slowly been disseminated to enthusiasts ever since. The only way to stand any chance of obtaining material is to discover who has it, then somehow inveigle a small piece of it your way via persistent and generous offers of trades. It resembles a compact form of *Ruscus aculeatus* but the

Ruscus aculeatus.

Ruscus hyrcanus.

cladodes are tightly arranged around the open-branched stems in spiralling whorls, reminiscent of a tiny little monkey puzzle tree. It is completely and utterly captivating. (H4/5)

Schinus

A small genus of aromatic trees of which one has an amount of hardiness. *Schinus molle* from arid regions in the northern parts of South America is a tall, weeping, evergreen tree with interestingly patterned, rough bark, elegant pinnate leaves and, following the insignificant flowers, strings of pink berries. All parts of the tree are strongly aromatic, the berries being the source of pink peppercorns. This is a popular ornamental tree in dry regions of the world being supremely drought-tolerant and providing light shade. I have grown this in my garden since 2015. By the end of most winters it is looking rather tatty. My late and early cold winters defoliated the plant and caused some tip dieback on the branches. It soon gets going, however, and makes a complete crown again by the end of the year. During hot summers I have had a small crop of pink peppercorns, although not quite enough for me to research recipes for pepper sauce. I am encouraged by this and feel it is a fabulous addition to the UK xeric garden, being so totally redolent of hotter, drier climates. So far so good. (H3/4?)

Schinus molle.

Schinus molle pink peppercorns!

Schinus molle leaf detail.

I grew *Schinus terebinthifolia*, which resembles a larger version of *Pistacia lentiscus* (and an even closer resemblance to *Pistacia terebinthus*, hence the name), but is less hardy. Relatively mild winters caused the leaves to be damaged and it was killed by my early cold winter. I would say it is worth a try if you have a very hot spot, as whenever I see this plant I think of the Canary Islands, where it is a very common ornamental plant. (H3)

Tamarix

Tamarisks are a group of drought-, salt- and wind-tolerant plants most often found planted in coastal areas as windbreaks, often right at the edge of the sea. From stout, rough-barked trunks they grow a hazy tangle of wiry branches covered in conifer-like scaly leaves and bear masses of small flowers along the branch tips. Several species are grown ornamentally, notably the pink-flowered *Tamarix tetrandra* and white-flowered *Tamarix africana*. As a general rule I find these a little too scruffy for gardens but along the seafront at Westcliff-on-Sea, Essex there are some amazing old plants of *Tamarix africana* that are regularly pollarded and, in full flower, look spectacular. For windy, coastal sites, these are the perfect plants to give you the wispy desert plant look, especially with a little judicious pruning. (H5)

Tamarix africana on the seafront at Westcliff-on-Sea.

HERBACEOUS

Individual plant entries are brief in this section, so rather than give a hardiness rating each time, assume H4/5 unless stated otherwise.

Acanthus

A fabulously architectural group of herbaceous perennials and shrubs that are important plants for the xeric garden. The first three form a natural group in terms of how they look. The leaves of *Acanthus mollis* have shallow lobes compared to the deeply cut, spiny-edged leaves of *Acanthus spinosus*, which, when taken to extreme in *Acanthus spinosus* Spinosissimus Group, become almost skeletally reduced and thistle-like. All grow in a similar fashion, producing clumping patches of arching leaves and erect, spiny spires of mauve-hooded, white flowers, although the *Spinosissimus* form is less floriferous. They are very easy to propagate by root cuttings, to which anyone who has tried to move one will testify, as they keep popping up in perpetuity from tiny scraps of roots left behind. Unfortunately *Acanthus mollis*, and to a lesser extent *Acanthus spinosus*, can be disfigured by a powdery mildew in hot weather when there is little air movement. There is not much to be done other than removing the affected leaves, but that can leave quite a gap. The next two species seem immune to this problem, thankfully. *Acanthus hungaricus* is smaller than the above group of species, with leaves that are quite deeply cut – more than *Acanthus mollis* but not so much as *Acanthus spinosus*. It makes a tidier, more compact plant with similar purple-hooded white flowers. *Acanthus hirsutus* is smaller and neater still, with hairy, green flowers and hoods. Once established it has a tendency to run a little bit underground but never so much as to be a problem. I find it especially attractive and it is my favourite of the purely herbaceous species.

The shrubby acanthus are very interesting, there potentially being a number of woody species from Africa that, in this country, are cut down by winter cold but return again either from wood or underground. *Acanthus sennii* has been grown for decades and, as a herbaceous plant, seems hardy to around –15°C. This will make 1.2m of shrubby growth annually with extremely prickly, holly-like leaves that are beautifully marked with prominent vein patterns. Late in the year (sometimes not quite making it before the frost), it produces spikes of bright red, hooded flowers that are simply spectacular. Even if the stems persist through winter it is best cut back in spring to encourage more compact growth. There are other species that have not been grown for as long but are just as interesting, including *Acanthus arboreus* and *Acanthus polystachyus* (syn. *Acanthus montanus*). These, if anything, have even more beautifully marked and shaped leaves with the most vicious, hard prickles on the margins. I am not sure if these would flower on the current season's growth but are worth growing for the foliage alone. *Acanthus arboreus* has been known to return herbaceously after

Acanthus mollis.

Acanthus spinosus Spinosissimus Group.

Acanthus spinosus.

Acanthus hungaricus.

Acanthus hirsutus.

Acanthus sennii.

Acanthus arboreus.

–5°C. *Acanthus polystachyus* is too new for this to be known but, if it comes back, it is massively exciting and potentially the best of all.

Agapanthus

Widespread in cultivation and extremely familiar, there are just so many species, hybrids and cultivars that I imagine it would be an easy thing to fall into the rabbit hole of collecting them. Fortunately I find the Headbourne hybrids satisfy my requirements, as they bulk up quickly, flower regularly just when you need that gorgeous cooling blue and self-seed generously. So far I have managed to resist the temptation to look for anything more out of the ordinary. There are, however, those lovely deep inky purple ones…

Asclepias

A huge genus of perennials of which I have recently started growing just two. *Asclepias curassavica* is a shrubby plant but rather frost-tender. Mild winters will see it cut to the ground but it should re-sprout. My early cold winter saw it off but it self-seeded and popped up in a few places. The flowers, produced in late summer and autumn, are a vibrant

Asclepias curassavica.

mixture of red and yellow and attractive to butterflies. *Asclepias tuberosa* is hardier, reliably returning, with bright orange flowers but not quite as showy.

Asphodelus

Asphodelus albus will be familiar to anyone who has been stuck behind a group of cyclists while driving through the Tramuntana Mountains in Mallorca in the spring. Or, rather, being driven: drivers need to keep their wits about them on those twisty roads. Here they flower a bit later, producing their spires of white, lightly-striped stars in early summer. *Asphodeline lutea* is similar, clothing its stems with grey-green bottle brushes of glaucous, chive-like foliage topped by bright yellow, starry flowers.

Ceratostigma

Ceratostigma has a couple of valuable, drought-tolerant deciduous shrubs, *Ceratostigma willmottianum* and *Ceratostigma griffithii*, that are ostensibly similar, but my favourite, by some measure, is the herbaceous/sub-shrubby *Ceratostigma plumbaginoides*. All three bear clusters of cobalt-blue flowers late in the season from deep-red bracts, by which time the leaves are turning red, making a warm show of colour. In contrast to the others, *Ceratostigma plumbaginoides* remains low, spreading from underground and making a drought-proof carpet – the hotter and drier the spot, the poorer and stonier the soil, the better it grows and the stronger the red colour. Allow it plenty of space to expand. It is a bit of a nuisance in spring when all of the dried stems need to be snipped off but no more so than, say, some hylotelephiums.

Cynara

The massive and familiar cardoon, *Cynara cardunculus*, makes a bold statement where there is room to do it justice. It comes into leaf over winter and flowers at up to 3m high in summer with grapefruit-sized thistle flowers on branched stems. In the worst droughts the leaves die back, so that only the towering, spiny, dried flower-stalks remain in a kind of surreal floral arrangement. In areas prone to blackfly it is probably best avoided as they can spoil the look of an entire plant in just a couple of weeks and are all but impossible to eradicate. Fairly new to me, but already usurping the ordinary cardoon, is *Cynara cardunculus* subsp. *flavescens*, which I find has far more of interest going on. This is half the size, so easier to accommodate. The silver leaves are much more deeply cut and divided, and the tips of each leaflet are reduced to sharp, golden spines. The flowering stems are covered in spikes and the fat bracts from which the bright violet florets extend are, you guessed, reduced to

Asphodelus albus.

Ceratostigma plumbaginoides surrounding and softening this agave.

Cynara cardunculus subsp. *flavescens*.

spines. The whole thing exudes spiky menace! I unexpectedly bumped into this several years ago in Lanzarote, where it was growing wild on the top of a cliff, and have sought it out ever since, eventually finding a supplier just a couple of years ago.

Dicliptera

A large genus of shrubs and herbaceous plants related to acanthus, *Dicliptera sericea* (syn. *Dicliptera suberecta*) reputed to be tender but mine came back vigorously after my early cold winter. Rounded, furry, grey leaves carry 60cm-high stems of whorled, showy, orange-red, trumpet-shaped flowers from summer until the frosts. Great for butterflies. I find it a useful repetition of colour for *Zauschneria californica* but in a different plant.

Dicliptera sericea.

Echium

There are many species of *Echium*, even a UK native *Echium vulgare*, but for me the most interesting come from the Canary Islands and, of these, there are three particularly rare ones (in the wild) that are most suitable for growing outside in the UK. The most familiar has to be *Echium pininana*, a towering triennial triumph of sapphire-blue flowers reaching 4m or more in my arid conditions and a good deal taller in the wetter west of the UK. They are, by some definitions, biennial, i.e. a 24-month life-cycle but spread across three calendar years. Plants are in flower for weeks from late spring to late summer, dropping seed that germinates in autumn and early winter to make a carpet of oval, hairy, speckled leaves. Throughout the second year these little seedlings grow quickly to become chest high on solitary trunks by the end of the growing season; the leaves losing their speckles but gaining size and coarseness. Then, in their third year, there is the spectacle of the rapidly expanding flower spike, the cylinder of flowers buzzing with bees throughout the summer. In the wild these are only found in the *laurisilva* forests on the island of La Palma, where I have seen them growing on damp, shady, canyon sides at the magical sites of Los Tilos and Cubo de Galga. Temperatures around –5°C or slightly lower will damage or kill plants but seed is much hardier: once dropped it can remain viable for several years afterwards. I have a white-flowered form that has stayed relatively true, seedlings in the same general area still growing up to be white-flowered fifteen or so years later. It is possible to tell them apart from the blue ones by the absence of the reddish stem colour that blue-flowered individuals have.

Echium wildpretii is a stunningly beautiful plant from Tenerife, growing from a basal rosette of narrow, hairy, silvery leaves, then pushing up an astonishing tall, rusty-red spire of flowers in its third year. These live almost exclusively at the top of the island above the treeline in the moonscape of Las Cañadas around the base of Mount Teide. If you go for a walk around Los Roques de Garcia in the third week of May, the 1.5m-high red beacons are absolutely everywhere, announcing their presence from the rocks they were formerly hiding behind. This species is a lot hardier to cold than *Echium pininana* but intolerant of wet feet and difficult to please. It is best grown in an inorganic substrate, as in nature it grows in degraded lava rock with no organic content at all. On occasion I have had both these species flowering in my garden at the same time and the ceaseless activity of bees ensures the subsequent few years' seedlings are always interesting hybrids, some favouring one parent in overall habit, some the other, but almost always with pink-amethyst flowers. The very best resemble a giant *Echium wildpretii*. There is a subspecies, *Echium wildpretii* subsp. *trichosiphon* that is endemic to La Palma, again to be found above the treeline at the very top of the island around the Roques de los Muchachos Mirador and observatories. It is very similar to the type species but with the same pink-amethyst-coloured flowers of my garden hybrids. This is interesting as my third favourite *Echium* species, *Echium gentianoides*, grows alongside it. This is a branched, shrubby species with thick, coarse, leathery leaves and, as the name implies, bright-gentian coloured flowers, borne on open panicles from the tips of the branches. Does this make *Echium*

Echium pininana, white and blue.

Echium wildpretii at the top of Tenerife.

Tall hybrid favouring *Echium wildpretii* in appearance.

Echium gentianoides.

Echium pininana: bees getting busy…

Echium candicans is fast growing but quite tender.

wildpretii subsp. *trichosiphon* a stable hybrid population? Anyway, I have found *Echium gentianoides* to be the hardiest of the Canarian echiums I have tried (which is most of them) and it makes a very attractive, if short-lived, small bush for the xeric garden. Like *Echium wildpretii*, this is best grown hard in a rocky, mineral substrate in full sun.

Eremurus

A large genus of herbaceous perennials that produces the most amazing dense, wind-firm spikes of flowers from a pretty nondescript tuft of grey-green leaves, the height and colour varying according to the species or cultivar. They grow from a fleshy, extremely brittle root system that resembles some kind of alien hunting spider and are difficult to please. They need freely draining soil but if planted too deep, the fleshy roots rot; too near the surface and they do not get going. They need open space around them so that in summer the roots can get a good unobstructed baking. I have tried many times to establish them and, until my last attempt, failed, but have now finally managed to get them growing. At RHS Hyde Hall they are spectacular, even self-seeding around the garden. *Eremurus robustus* can have flowering

Eremurus 'Cleopatra'.

stems reaching 3m tall with white flowers, while those of *Eremurus stenophyllus* reach around 1m tall and are golden yellow. The cultivar 'Cleopatra' has warm peachy-orange flowers and reaches 1.5m.

Eryngium

This is a massive genus of plants from all over the world, including our UK native sea holly, *Eryngium maritimum*, that is to be found growing on sand dunes along our coasts. There are so many that grow well in poor, stony, dry situations that it is hard to narrow down to just a few, so I will choose just one that I favour above all others. *Eryngium bourgatii* 'Picos Blue' is a stunning plant, from the ruffled edges of its finely cut, beautifully marked leaves to the intense metallic blue of the spiked collar on which the thimble-shaped thistle flower sits. Planted *en masse* it makes

Eryngium maritimum growing in dunes at Shoeburyness.

Eryngium bourgatii 'Picos Blue'.

the most spectacularly textured carpet when grown underneath taller, spiky plants. There is a sort of hotspot for *Eryngium* in the Americas. Some of these plants are extremely exotic-looking with names that give clues to their appearance, *Eryngium yuccifolium* from the USA, *Eryngium pandanifolium* from Brazil down to Argentina, *Eryngium agavifolium* from Argentina and *Eryngium bromeliifolium* from Mexico. I have found all these to need a wetter climate than mine, however, plants dwindling away without extra irrigation when grown in perfectly draining soil. This is especially so for *Eryngium pandanifolium*, which will most happily grow in a pond margin; this is actually a good way to grow it as it prevents the leaf bases from becoming a snail shelter, a very real risk when grown on dry land. Where these plants excel is providing an interface between the xeric garden and the rest of the exotic garden. They represent a kind of half-way house in terms of looks and growing requirements, alongside other spiky, toothy and succulent-looking plants, such as *Morina longifolia* and pokers such as *Kniphofia northiae* and *Kniphofia caulescens*.

Eschscholzia

A genus of a dozen annual or short-lived perennial poppies from the USA, only one of which seems to be in cultivation in the UK for reasons I intend to research. *Eschscholzia californica* is a really handy plant for providing easy-care, space-filling colour in the summer. It could not be easier – in autumn scatter seeds on the surface of where you want it to grow then forget about it. Seed will usually germinate through winter and into spring with the first plants starting to flower in April. Once they get into their stride in May and June they will be visible from space and latecomers will still be flowering into September, although the scruffier plants are best removed by then. In the course of removing the old stems, seeds for next year will fall from the disturbed pods and so the cycle continues. If they germinate and start growing where you

Eschscholzia californica.

Ferula communis.

Glaucium flavum.

Glaucium corniculatum.

do not want them, the deep taproots pull out very easily. Various colours have been selected, from pale primrose through the usual orange into bright red: sometimes a solid colour, sometimes shaded within the petals. For sheer eye-watering value, however, the orange is hard to beat.

Ferula

Related to true fennel, *Ferula* comprises over 200 species that typically grow in dry, open, rocky places across Eurasia and beyond. *Ferula communis* is a wonderfully architectural plant. Frothy piles of finely cut, fresh green leaves start to be produced in late winter. After a few years the plant has established enough to flower and produces a colossal branched spike of pompom umbells, containing thousands of tiny yellow flowers, reaching over 3m high in May/June. The leaves then die back as the heat of summer triggers dormancy, leaving the huge flowering stems behind. *Ferula communis* subsp. *glauca* is extra special with new growth that is flushed purple as it emerges, ageing to glaucous grey. Flowering stems are tinted purple and flowers are a touch darker. These need space as they will spread to 90cm or so, and it is also best if their summer-growing neighbours do not entirely shade the roots, which need baking. It is said that if a plant starts to flower prematurely, in its first or second season, it is at risk of dying prematurely afterwards, so it is best to cut out the flowering stems before they reach full size to encourage the plant to grow as a perennial. I have not had to do this but keep the idea in reserve. Plants may not flower every year but, even just with their foliage, they make a marvellous contribution to the xeric garden, as the soft, billowing leaves provide such a great contrast in texture to the spiky things. Fairly new to me is *Ferula lancerotensis*, which, as the name suggests, is from Lanzarote. This is another summer-dormant plant that has, if anything, more finely dissected leaves than *Ferula communis*, albeit smaller overall, but similar yellow pompom flowers. My plant has been in the ground for three years and has grown through my early cold winter without too many problems, taking down to –5°C.

Glaucium

A small genus of perennial xeric poppies, one species of which is native to the sand dunes and shingle beaches of the coastal UK. *Glaucium flavum*, which absolutely thrives in hot, sunny, rocky conditions, has wonderfully crinkly and lobed, waxy, silvery leaves and pretty yellow poppy flowers in summer. Flowering is followed by the production of extremely long, narrow seed pods, leading to it being given the common name horned poppy. The whole plant is extremely toxic, so handle it with care! *Glaucium corniculatum* looks very similar but with bright orange-red flowers and is also toxic. Both are worthy of a place in the garden and, to be fair, there are plenty of other toxic plants we routinely grow in our gardens that barely get a mention.

Helleborus

A genus of extremely useful evergreen perennial plants that usually flower in late winter to early spring. These Lenten and Christmas roses are fabulous plants for woodland-type gardens but a couple of species have a spikiness about the leaf structure that makes them sit happily with xeric plants. *Helleborus argutifolius*, from Sardinia and Corsica, is a fabulous plant producing upright sprays of pale green, wide, saucer-shaped flowers. Foliage is leathery, grey-green and finely-toothed along the margins. *Helleborus lividus* comes from Mallorca and is even more beautifully marked but, sadly, it is substantially more tender. I have never managed to keep it alive, despite numerous attempts. There is a hybrid between these two, *Helleborus × sternii*, that for me combines the best of both parents and is a superlative plant, with the extra colour from *Helleborus lividus* but the size and hardiness of *Helleborus argutifolius*. The cultivars 'Boughton Beauty' and 'Silver Dollar' are good named selections from this cross. I find these sit well with larger succulents, such as yucca, nolina and dasylirion, but are especially useful, as with the eryngiums mentioned earlier, when trying to move visually from an arid border to one containing more general exotic-style planting.

Helleborus × sternii.

Helleborus argutifolius makes effective ground cover around large, spiky plants.

Kniphofia

A genus of perennials native to Africa that is in the same family as aloes, with flower spikes that are very similar. We can once again use this resemblance to our advantage when moving away from the xeric borders but repeating the same shapes and colours of the flower spikes. There are many, many species suitable for this purpose but for me two stand out. *Kniphofia caulescens* is a low, spreading plant with angular, stiff-looking but soft, glaucous, blue-grey leaves. It has marvellous two-tone strawberries-and-cream flower spikes that change from pinky-red to off-white as they age, so that spikes are strawberry at the top, cream underneath. The most impressive of all has to be *Kniphofia northiae* but it seems difficult to find accurately named plants currently, there being many similar, but not quite as impressive, hybrid imposters around. The real deal is a huge plant with massive leaves, over 1.2m long, up to 12cm wide

Kniphofia caulescens at RHS Hyde Hall.

Kniphofia northiae imposter.

Old picture of the real *Kniphofia northiae*.

at the base and rounded in cross-section with no keel. A densely packed, chubby poker of brownish-red fading to cream flowers is held more closely to the leaves than most. It bears a startling resemblance to *Aloe speciosa*, even to the point of growing a short trunk. Neither is happy in xeric growing conditions but absolutely perfect for leading the eye away from the spiky plants to the larger exotics.

Lewisia

A genus of flowering plants from rocky places in the western USA. They are perfect for planting in east- and north-facing crevices in rock gardens for which we would otherwise struggle to find succulents to grow. Arching sprays of colourful flowers are produced in late spring to

Lewisia cotyledon cv.

early summer, and plants live for many years, slowly making a small stem. There are many species, *Lewisia cotyledon* being one of the more popular ones, but most garden centres will carry any number of hybrids and cultivars in their alpine plant sections. I must admit that I have never looked into the pros and cons of the various species but simply chosen the variety I want to grow by the flower colour. The most important thing is that their fleshy roots are given perfect drainage; if possible they should be grown wedged into a vertical rock face.

Lomandra

Lomandra longifolia is another plant for the spiky/exotic interface. It is a tough evergreen perennial from eastern Australia and Tasmania that grows into a large clump of narrow, strappy leaves resembling a chunky grass (although they are actually closer to lilies). Established clumps throw up curious flowers like spiky clubs that nestle among the leaves. Eventually they mound themselves up to form a kind of trunk: I have seen an impressively large specimen in the Botanical Gardens in Porto, Portugal, that was nearly 2m high. They will grow in dry, but not completely xeric, soil and form a good textural contrast for those edges near the xeric border.

Lomandra longifolia at Tremenheere Sculpture Gardens.

Pelargonium

A huge genus with lots of interesting plants. The one I love, and which is doing well against expectations, is *Pelargonium sidoides*. This is buried deep inside a crevice in my rock

Pelargonium sidoides.

garden and each year, if cut back by frost, puts out wiry stems of scallop-edged, heart-shaped leaves that are covered in silver fur. Wispy sprays of deep mauve, almost black, flowers hover about the plant all summer. This plant needs absolute drainage and full sun: given those, mine has survived my early and late cold winters.

Penstemon

Another huge genus of dry garden favourites from, mainly, North America. *Penstemon barbatus* gets the mention here for its upright spikes of tubular, vivid scarlet flowers

Penstemon barbatus.

with hairy lips. Grown very hard in poor, rocky soil and full sun, it is more inclined to behave itself and not flop over. In some ways it does a similar job to *Verbena bonariensis* in that it produces a transparent haze of scarlet, rather than mauve, that seems to float above its companions.

Perovskia

Now considered part of the genus *Salvia*, one species is commonly grown and widely available. *Perovskia atriplicifolia* (more correctly *Salvia yangii*) is one of the bastions of the dry garden, providing reliable colour and texture, even when there is absolutely no rain at all during summer. Plants are best cut back close to the base each spring from where they make an erect shrub to around 1m tall that bears small, lilac, hooded flowers all summer and into autumn. The cultivar 'Blue Spire' is the easiest to find.

Salvia

A massive genus of over 1,000 species, many of which are perfectly suited to the xeric garden. I am going to choose just two to mention.

Salvia confertiflora

This has a different look to most salvias. The olive-green leaves are quite large, 20cm or so long, and beautifully 'sage'-scented. The flowers are also very unusual: short, bright orange-red tubes borne in whorls along an upright, deep-red spike. In poor, stony soil in full sun this will make a 1.5m-tall, branched plant in one season from cuttings, but it is not particularly hardy. My plant came back from –2°C but was killed at –5°C.

Salvia elegans

Known as the pineapple sage, it is so-called because the foliage smells strongly of pineapple when stroked. Very late in the year, October or November, almost luminous scarlet flowers are held in profusion above the leaves on branched stems. I have not grown it for that long but in a mild year it continued to flower

Salvia confertiflora.

until mid-January. In spring, cut back to ground level and the plant comes back from the roots. This is unsurpassed for eye-catching late colour.

Saxifraga

A large genus, many of the species of which grow in rock crevices at high altitudes. There are books dedicated to them so I will not go into any detail, only mention that they exist and ought to be considered. Most garden centres have a good selection in the alpine plant section and they are just

Saxifrage cv.

the perfect little plants to fill in cracks between succulent plants in the rock garden. I especially like the tight domes of the silver- or crusty-leaf saxifrages, with leaves that appear to be edged in frost.

Sedum

Actually none of these are still sedums but most people would look for them under this name. There are so many, here are just a few favourites. *Hylotelephium* 'Bertram Anderson' is a trailing, deciduous plant with glaucous-grey leaves and a tinge of purple that is enhanced as the light intensity increases. Bright pink, starry flowers cover the plant in late summer. *Hylotelephium* 'Vera Jameson' is similar but slightly more erect, with darker foliage. *Hylotelephium cauticola* 'Lidakense' makes wonderful low, spreading mats of tiny glaucous-grey rosettes with just a hint of purple. *Hylotelephium* (Herbstfreude Group) 'Herbstfreude' is a horrible mouthful for what we used to call *Sedum* 'Autumn Joy' in the old days. For my money this is still the best of the pale-leaved *Sedum spectabile* forms. *Hylotelephium* 'Matrona' is slightly more erect and chunky, and has the most wonderful three-dimensional purple overlays on the fat, glaucous grey leaves. *Hylotelephium telephium* (Atropurpureum Group) 'Karfunkelstein' is another memory-testing name for what I consider to be the best of the purples. This has a strong navy-blue quality that is difficult to describe, and even more difficult to photograph, remaining on the blue side of purple and not edging into chocolate as some do. *Hylotelephium populifolium* is quite different, with lax, wiry stems of pale-green, fleshy leaves that have deeply toothed margins. Clouds of white flowers are produced in late summer. This actually self-seeds, although it is not a nui-

Hylotelephium telephium (Atropurpureum Group) 'Karfunkelstein' still colourful at the end of the year.

Phedimus aizoon.

Rhodiola pachyclados.

Rhodiola rosea.

sance. *Hylotelephium sieboldii* is gorgeous: trailing stems of glaucous-grey, silver-dollar foliage that takes on a red edge in good light. Masses of pale pink flowers are produced towards the end of summer, by which time the leaves can start to turn reddish before falling. *Phedimus aizoon* is a colourful plant early in the year with fresh, apple-green leaves on upright, reddish stems, all of which contrast nicely with flat heads of bright yellow, starry flowers. It can, however, look a bit tatty later in the year. *Rhodiola pachyclados* makes useful ground cover: spreading mats of china-blue rosettes comprising notch-tipped leaves. I have found that it struggles in dry years, so further west it would be superb. Lastly, *Rhodiola rosea* is a great thing: glaucous-grey whorls of leaves around unbranched stems topped by a dome of yellow, starry flowers, the overall effect being rather like a deciduous *Euphorbia myrsinites* but growing in a neat, tight bun.

Thapsia

Thapsia garganica has finely cut, tough-textured, glaucous-grey leaves and, in summer, umbells of yellow flowers. It comes from rocky, open places in the Mediterranean area and is extremely drought-tolerant. The colour and texture of the leaves make it a little different to the usual Apiaceae but, frankly, a plant

Thapsia garganica pompom.

Thapsia garganica leaf detail.

with the common name of 'deathly carrot' has got to be worth growing for that reason alone. The roots are mildly toxic and careless handling is said to cause skin rashes. It comes into growth early, as do a lot of the carrot family, and mine has taken –5°C with no issues.

Tinantia

A handful of species from the Americas closely allied to *Tradescantia*. *Tinantia pringlei*, from high mountains in north-east Mexico, is a cute thing with green 'wandering Jew' leaves that are spotted with purple – completely flushed purple in strong light – and cute little pale-mauve, three-petalled flowers. Cutting back immediately after the first wave of flowers curbs its fondness for self-seeding and keeps the plant tidy, encouraging a repeat flower show later in the year.

Tinantia pringlei.

Grown on poor stony soil in full sun it is absolutely gorgeous.

Verbena

Verbena bonariensis became a very fashionable plant with the advent of the 'gravel garden', championed by the likes of Beth Chatto and John Brookes in the 1970s and 1980s. Unlike some fad plants, this one has stood the test of time and still remains as popular now as ever. For a good reason – few plants do what this does so well. It is completely drought-tolerant. In fact, the stonier and poorer the soil, the better this grows, remaining tighter and stronger, albeit shorter, and more reliably wind-firm. It then flowers non-stop from early summer until the first heavy frosts. But it is the way in which it flowers that makes this such a special plant. The flowers are borne on

Verbena bonariensis.

Verbena rigida.

slender but stiff, square stems, appearing as a tight cluster of slim, tubular, purple to mauve flowers that appear to hover over the foliage. In a peculiar optical illusion it produces colour and occupies spaces yet you can see through to the background – an extremely useful attribute if you want to have this colour but not have one of the blockier plants, such as Russian sage or lavender. Another useful species is *Verbena rigida*, which carries almost identical flowers but on a ground-hugging, carpeting perennial.

Zauschneria

Zauschneria californica is more properly now called *Epilobium canum* but most people still know it by the old name. It is a low-growing, shrubby perennial from the chaparral of California with silvery-

Zauschneria californica 'Ed Carman' surrounding *Agave gentryi*.

grey leaves and, from late summer until the frosts, bears a succession of tubular, bright red flowers. Fluffy seeds emerge from capsules in autumn, adding to the charm of this useful plant. Plants spread from underground and are best tidied back in spring, although a position in full sun and poor rocky soil helps keep them compact. Several different cultivars have been selected that vary in how silver the leaves are, overall growth habit and flower colour. 'Glasnevin' (syn. 'Dublin') has the typical grey-green leaves and bright flowers. 'Ed Carman' has greyer leaves and flowers on the orange side of red. 'Western Hills' has flowers of a stronger red and grows more upright with narrower, greener foliage. 'Olbrich Silver' has the most silver foliage of all and strong red flowers. Flip a coin.

GRASSES, BULBS AND FERNS

Grasses

The billowing softness of many grasses makes them the absolute perfect foil for the rugged architecture of spiky plants; it is just a question of scale when matching them as companions. The rarely seen *Ampelodesmos mauritanicus* gets my vote to represent larger grasses. It gives the same effect as plants such as *Stipa gigantea* or *Cortaderia selloana* but, in my opinion, does it with more style. Piles of deep green, glossy, evergreen leaves are topped by towering, soft, oat-like plumes that stay in good condition through most of the winter. Considering it comes from Mallorca it is surprisingly hardy, there being long-established clumps at RHS Hyde Hall. Unsurprisingly, it is totally drought-proof.

Moving down in scale, *Festuca glauca* is an extremely popular grass with tussocks of fine, glaucous, blue-grey leaves. The hotter, drier and sunnier the spot, the more intense the colour becomes. Various cultivars have been selected that show a stronger blue colour and include 'Elijah Blue', 'Intense Blue' and 'Blue Fox'. A haze of blue-green flowers,

Ampelodesmos mauritanicus.

Pennisetum villosum.

fading to buff, is produced in summer but they do not add value and I generally cut them off. Plants are best combed through over winter to remove dead leaves, otherwise the tussocks can look a bit scruffy. When it comes to billowy soft grasses, few can rival *Pennisetum villosum* and *Pennisetum alopecuroides*, both of which have the 'dangly fluffy caterpillar' look about their flowers that is so enchanting. Most hot places where succulents are found have a native one of these of one sort or another, so it is another one of those handy little subliminal mind-tricks to use a hardy look-alike. I have the glaucous blue-grey form of *Pennisetum macrourum*, which is a larger and stiffer plant than the last two, reaching 1.5m in flower, at which point the whole plant resembles an explosion in a firework factory. It is very tough and will slowly expand from the original clump via short, underground rhizomes, while also seeding around. Given the

Pennisetum macrourum, glaucous form.

space, a plant growing in full sun on poor, stony soil is a wondrous sight to behold. *Stipa tenuissima* (now more correctly *Nassella tenuissima*) is another widely grown grass that can combine spectacularly with some of the larger succulents. A simplified planting scheme comprising a huge drift of this growing around a few specimens of *Dasylirion acrotrichum* or *Nolina nelsonii* would look totally stunning. Sadly I lack the self-restraint to do this for myself.

Bulbs

Bulbs offer the opportunity to provide passing seasonal colour within a framework of architecturally strong but chromatically static plants. Some are great for drifts of colour, some are great to zoom in upon and enjoy the detail. *Amaryllis belladonna* is a fabulous bulb to grow for late season colour. Fragrant pink trumpets are borne on naked reddish stems, which die down just as the leaves appear. They must not be buried deeply and it is often said that it is best to keep the 'neck' of the bulb at ground level so that the bulb can get a good baking in summer. The leaves take up more space than is often anticipated, so need to be taken into account when considering neighbouring plants. Also be aware that they are toxic. *Nerine bowdenii* grows in a similar cycle, producing its spidery, candy-floss-pink flowers on naked stems in autumn. Again, it is best planted close to the surface in a hot, sunny position. Many

Crocus species are native to hot, open, dry slopes so enjoy the same conditions as many of our succulent plants. It makes a welcome change from seeing them planted in drifts in lawns. There is an endless labyrinth of *Crocus* to be explored but for the more casual grower, like me, most bulb catalogues contain what we need for that blast of early colour without paying a hefty price for provenance or worries that the badgers will eat them. Much the same could be said for *Cyclamen*. I felt it necessary to include them as they are so valuable in rock crevices for so many reasons, but I do not really know much about them. I have a rough notion in my mind that *Cyclamen hederifolium* flowers in autumn, whilst *Cyclamen coum* flowers in spring, beyond which my knowledge rapidly falls off. These two really just scratch the surface, there being plenty of other species to enjoy. I find one of the most rewarding aspects of cyclamen is the almost infinite variety of leaf patterns, almost like miniature tapestries in the complexity of their shading, veining and marking, even within the common species.

Amaryllis belladonna.

Cyclamen leaf patterns.

Fritillaria persica is my pick from this large genus of bulbs; a native of rocky slopes across the Middle East it is happy growing alongside our succulents. Waking up early in the year, it produces tall spires of deepest maroon, almost black, bell flowers clothed at their base with glaucous-grey foliage. It is a source of frustration for me that every time I plant these, my badgers seem to find them the same night, dig up the bulbs and eat them. I will keep trying with increasingly robust barricades. For a jolt of bright colour *Gladiolus communis* subsp. *byzantinus* is hard to beat. I have recently planted quite a few in an area populated with *Eschscholzia californica*: the timing is good, the contrast spectacular. For adding a little interesting detail, *Iris tuberosa* (syn. *Hermodactylus tuberosus*) is a great choice with its early, green and black flowers. In appearance it is much the same as *Iris reticulata*, another early bulb with a selection of colours available. Both naturalise well. As will *Muscari armeniacum*, which will take hold of the floral baton and flower once the above have finished and provide a good show of those cute, bobbly, blue flowers. *Scilla peruviana* is a huge favourite of mine. Wide, strappy leaves are produced in autumn and carry on growing through winter and spring. In early summer, outrageously big conical heads of starry, lilac-blue flowers are produced, after which the whole thing disappears for summer, when it appreciates a baking. Another gem, *Sternbergia lutea*, is, in effect, an autumn-flowering crocus. Leaves appear in early autumn, followed by the beautiful clear golden-yellow, crocus-like flowers, the leaves remaining green until spring when the plant retreats below ground. This is another plant that enjoys a baking in summer. As do most of the dwarf species of tulips, of which there are legion. My favourite, because it stays so compact, is *Tulipa linifolia*. Grey-green leaves, some with cute wavy margins, spread across the ground and stumpy stems, just 15cm high, bear bold scarlet flowers with a black throat in April. Lastly, *Drimia maritima* (syn. *Urginea maritima*) is something of a diva and needs to have some space of its own. Wide, fleshy leaves are produced in autumn from a huge onion-like bulb, which die down in summer. Once established, usually after several years, an

Nerine bowdenii.

Tulipa linifolia.

Gladiolus communis subsp. *byzantinus* with backdrop of poppies.

Drimia maritima.

immense spike comprising hundreds of starry, white flowers is produced in late summer, seemingly out of nowhere. This species comes from coastal and rocky areas around the Mediterranean and, when growing in the UK, needs absolute drainage, hot summers and relatively mild winters; but if you can, you should!

Ferns

It may seem counter-intuitive but there are several ferns that enjoy the same conditions as cacti and succulents. If you visit any xeric plant habitat there will almost certainly be ferns tucked away in the crevices between rocks, curling in on themselves during periods of drought, only to expand when rains appear again. I find these totally fascinating; they add such a great contrast of texture but also the whole 'these are ferns why are they growing here?' aspect to a xeric garden. There are some UK native ferns that fit the bill too, so not much of a risk when it comes to the climate and the odd bad winter. I have been growing *Asplenium trichomanes* in a north-facing crevice in my rock garden for something like twenty years, in which time it has seeded itself (spored itself?) into nearby – and not so nearby – crevices to make a thriving colony. *Asplenium ceterach* has established well and, although it looks worryingly as if it is at Death's door in summer, it has never failed to regenerate. This has also self-seeded, albeit not with the same vigour, but to a spot some 5m away, which I consider to be a pretty good effort. I have a selection of the native polypodies – *Polypodium vulgare*, *Polypodium cambricum* and *Polypodium interjectum*. These all disappear completely for summer, shedding their fronds as soon as the drought kicks in, but return regularly as clockwork with the rain in autumn, and keep their fronds all through winter. Again, these have started to appear in crevices where I have not planted them, so must be happy, although I could not tell you which of the species each new plant is.

Asplenium trichomanes.

Asplenium ceterach.

Polypodium of some sort.

The champion long-distance volunteer in my garden has to be *Asplenium adiantum-nigrum*, however. About ten years ago I noticed an unfamiliar fern growing behind one of my agaves which, it transpires, is *Asplenium adiantum-nigrum*, the black spleenwort. According to a fern enthusiast friend, the nearest plant of which he is aware is some 5km or so further east from here, growing in the walls of the old library in Leigh-on-Sea. I feel especially pleased to have such a visitor in my xeric beds.

From further afield (beyond the UK) I have a handful of *Cheilanthes* species growing here, of which *Cheilanthes tomentosa* is the easiest to find for sale. Their slightly hairy, grey-green fronds look fabulous all year round but especially when the white furry croziers first start unravelling. *Cheilanthes argentea* is less frequently encountered but very pretty, bearing beautifully dissected triangular leaves with a silver reverse. *Pellaea atropurpurea* is another one I have had here for many years, seeding around and setting up new colonies some distance away from the mother plant. There are plenty of species in both genera that are worth growing, the main problem is finding a source for them.

Astrolepis sinuata is another USA/Mexican plant that has long, narrow, upright, grey-green fronds, interestingly cut along the margins, with a gorgeous cinnamon furry reverse. *Polypodium scouleri* is an absolutely classy fern. From the coasts of western USA this really is exceptionally drought-tolerant and has the most wonderfully textured, thick leathery fronds. These non-native xeric ferns are best grown with their roots deep inside shady crevices but their leaves out in the sun.

Finally, *Phlebodium pseudoaureum* deserves a mention. It is commonly sold as a house plant but I have found it to be quite tough, taking my late and early cold winters in its stride; so, encouraged by this, I now have it planted in a few places. Initially I planted it deep into a crevice in my rock garden from where the furry rhizome has crept along between and over the rocks. The pale, glaucous blue fronds really do stand out.

Cheilanthes tomentosa.

Cheilanthes argentea.

Pellaea atropurpurea.

Astrolepis sinuata.

Polypodium scouleri.

Phlebodium pseudoaureum growing in a drainage tube.

Shady rockery plants.

BIBLIOGRAPHY AND SELECTED FURTHER READING

Amidon, Jane. *Radical Landscapes: Re-Inventing Outdoor Space* (Thames and Hudson, 2001)

Bogler, David John. *Taxonomy and Phylogeny of Dasylirion* (The University of Texas, 1994)

Brookes, John. *Garden Design Workbook* (Dorling Kindersley, 1994)

Carter, S. *et al*. Aloes:*The Definitive Guide* (Royal Botanic Gardens, Kew, 2011)

Chance, Leo J. *Cacti and Succulents for Cold Climates* (Timber Press, 2012)

Chatto, Beth. *The Dry Garden* (Dent, 1978)

Chatto, Beth. *Beth Chatto's Gravel Garden* (Frances Lincoln, 2000)

Gentry, H.S. *Agave of Continental North America* (Arizona, 1982)

Green, Cheryl W. and Ferguson, Dave J. *Prickly Pears* (Cheryl Wright Green, 2005)

Hammer, S *et al*. *Mesembs of the World* (Briza Publications, 1998)

Hawker, Jon L. *Agaves, Yuccas and their Kin* (Texas Tech University Press, 2016)

Hochstätter, Fritz. *The Genus* Nolina (Speciale Piante Grasse, 2010)

Hochstätter, Fritz. *Yucca Vol. I, II and III*. (Fritz Hochstätter, 2000, 2002, 2004)

Horvath, Brent. *The Plant Lover's Guide to Sedums* (Timber Press, 2014)

Irish, Mary and Irish, Gary. *Agaves, Yuccas and Related Plants: a Gardeners' Guide* (Timber Press, 2000)

Jones, David L. *Cycads of the World* (Reed Books, 1993)

Jones, David L. *Palms Throughout The World* (Reed Books, 1995)

Latimer, Hugo. *The Mediterranean Gardener* (Frances Lincoln, 2001)

Lodé, Joël. *Succulent Plants of the Canary Islands* (Turquesa, 2010)

Nold, Robert. *High and Dry: Gardening with Cold-Hardy Dryland Plants* (Timber Press, 2008)

Olsen, Sue. *Encyclopedia of Garden Ferns* (Timber Press, 2007)

Pilbeam, John. *The Genus Echeveria* (British Cactus and Succulent Society 2008)

Pilbeam, John and Bowdery, Derek. *Ferocactus* (British Cactus and Succulent Society, 2005)

Pilbeam, John and Weightman, Bill. *Ariocarpus et cetera* (British Cactus and Succulent Society, 2006)

Reynolds, Gilbert W. *The Aloes of South Africa* (Trustees of the Aloes of South Africa Book Fund, 1950)

Rickard, Martin. *Ferns for a Cool Temperate Climate* (The Crowood Press, 2021)

Riffle, Robert Lee and Craft, Paul. *An Encyclopedia of Cultivated Palms* (Timber Press, 2003)

Schulz, Rudolf. *Aeonium: in Habitat and Cultivation* (Schultz Publishing, 2007)

Sierer, David. *Where the Hardy Cactus Grow* (The Amateur Digest, 1999)

Silver, Johanna. *The Bold Dry Garden: Lesson from the Ruth Bancroft Garden* (Timber Press, 2016)

Smith, Colin. *Yuccas: Giants Among the Lilies* (NCCPG, 2004)

Spain, John. *Growing Winter Hardy Cacti* (Elizabeth Harmon, 1994)

Spath, Jeremy and Moore, Jeff. *Agaves: Species, Cultivars and Hybrids* (Jeff Moore, 2021)

Starr, Greg. *Agaves: Living Sculptures for Landscapes and Containers* (Timber Press, 2012)

Steens, Andrew. *Bromeliads for the Contemporary Garden* (Godwit, 2003)

Stephenson, Ray. *Sedum: Cultivated Stone-crops* (Timber Press, 1994)

Taylor, Jane. *Plants for Dry Gardens* (Frances Lincoln, 1993)

The Cactus and Succulent Review. This is a free quarterly online magazine appealing to anyone interested in cacti and succulents for which I have written a number of articles: www.cactusandsucculentreview.org.uk

INDEX

First published in 2024 by
The Crowood Press Ltd
Ramsbury, Marlborough
Wiltshire SN8 2HR

enquiries@crowood.com
www.crowood.com

British Library Cataloguing-in-Publication Data
A catalogue record for this book is available from the British Library.

ISBN 978 0 7198 4416 4

Typeset by SJmagic DESIGN SERVICES, India
Cover design by Sergey Tsvetkov
Printed and bound in India by Nutech Print Services - India

Dedication

I dedicate this book to my wife Vicky, for cheerfully tolerating the relentless hijacking of our holidays by plants and, principally, for her endless support and indulgence which underpin my life.

Unless otherwise noted below, all pictures are taken by me of plants growing in my garden.

Acknowledgements

First and foremost I would like to thank Sheila Cude for proofreading the text before I sent it off to the publishers, painstakingly translating it from my flawed, conversational Essex prose into correct English. I'd like to thank the following for allowing me to use their photographs: Greg Starr, Daniel Marteeny, Tom Hart Dyke, Darren Hobbs, James Knock, James Tucker, Jan Kolendo, Melissa Scott, Mike Cullen, Pia Leppänen, Rob Stacewicz, Wayne Meyers and the late David Matzdorf. Finally, I'd like to thank Crowood for giving me the opportunity to write this book and for all the help and support they have given me throughout the process.

Image credits

Daniel Marteeny, p.123 (top and bottom left); Darren Hobbs, p.25 (bottom left); David Matzdorf, p.55 (top right); Greg Starr, p.76, p.81 (bottom); James Knock, p.56 (bottom); James Tucker, p.23 (bottom right); Melissa Scott, p.25 (top left); Mike Cullen, p.59 (left); Pia Leppänen, p.67 (top left); Rob Stacewicz, p.55 (bottom); Tom Hart Dyke, p.27 (bottom), p.110